PENGUIN BOOKS

YOUR INNER FISH

Neil Shubin is a palaeontologist in the great tradition of his mentors, Ernst Mayr and Stephen Jay Gould. He has discovered fossils around the world that have changed the way we think about many of the key transitions in evolution and has pioneered a new synthesis of expeditionary palaeontology, developmental genetics and genomics. He trained at Columbia, Harvard and Berkeley and is currently Chairman of the Department of Anatomy at the University of Chicago.

D0431562

YOUR INNER FISH

THE AMAZING DISCOVERY
OF OUR 375-MILLION-YEAR-OLD
ANCESTOR

NEIL SHUBIN

PENGUIN BOOKS

PENGUIN BOOKS

Published by the Penguin Group
Penguin Books Ltd, 80 Strand, London WC2R 0RL, England
Penguin Group (USA) Inc., 375 Hudson Street, New York, New York 10014, USA
Penguin Group (Canada), 90 Eglinton Avenue East, Suite 700, Toronto, Ontario, Canada M4P 2Y3
(a division of Pearson Penguin Canada Inc.)
Penguin Ireland, 25 St Stephen's Green, Dublin 2, Ireland
(a division of Penguin Books Ltd)
Penguin Group (Australia), 250 Camberwell Road,
Camberwell, Victoria 3124, Australia (a division of Pearson Australia Group Pty Ltd)
Penguin Books India Pvt Ltd, 11 Community Centre,
Panchsheel Park, New Delhi – 110 017, India
Penguin Group (NZ), 67 Apollo Drive, Rosedale, North Shore 0632, New Zealand
(a division of Pearson New Zealand Ltd)
Penguin Books (South Africa) (Pty) Ltd, 24 Sturdee Avenue, Rosebank, Johannesburg
2196, South Africa

Penguin Books Ltd, Registered Offices: 80 Strand, London WC2R 0RL, England

www.penguin.com

First published in the United States of America by Pantheon Books, a division of Random
House, Inc., 2007
First published in Great Britain by Allen Lane 2008
Published in Penguin Books with a new Afterword 2009
1

978–0–141–02758–6

www.greenpenguin.co.uk

Penguin Books is committed to a sustainable future
for our business, our readers and our planet.
The book in your hands is made from paper
certified by the Forest Stewardship Council.

TO MICHELE

CONTENTS

This book grew out of an extraordinary circumstance in my life. On account of faculty departures, I ended up directing the human anatomy course at the medical school of the University of Chicago. Anatomy is the course during which nervous first-year medical students dissect human cadavers while learning the names and organization of most of the organs, holes, nerves, and vessels in the body. This is their grand entrance to the world of medicine, a formative experience on their path to becoming physicians. At first glance, you couldn't have imagined a worse candidate for the job of training the next generation of doctors: I'm a paleontologist who has spent most of his career working on fish.

It turns out that being a paleontologist is a huge advantage in teaching human anatomy. Why? The best road maps to human bodies lie in the bodies of other animals. The simplest way to teach students the nerves in the human head is to show them the state of affairs in sharks. The easiest road map to their limbs lies in fish. Reptiles are a real help with the structure of the brain. The reason is that *the bodies of these creatures are often simpler versions of ours.*

During the summer of my second year leading the course, working in the Arctic, my colleagues and I discovered fossil fish that gave us powerful new insights into the invasion of land by fish over 375 million years ago. That discovery and my foray into teaching human anatomy led me to explore a profound connection. That exploration became this book.

YOUR INNER FISH

CHAPTER ONE

FINDING YOUR INNER FISH

Typical summers of my adult life are spent in snow and sleet, cracking rocks on cliffs well north of the Arctic Circle. Most of the time I freeze, get blisters, and find absolutely nothing. But if I have any luck, I find ancient fish bones. That may not sound like buried treasure to most people, but to me it is more valuable than gold.

Ancient fish bones can be a path to knowledge about who we are and how we got that way. We learn about our own bodies in seemingly bizarre places, ranging from the fossils of worms and fish recovered from rocks from around the world to the DNA in virtually every animal alive on earth today. But that does not explain my confidence about why skeletal remains from the past—and the remains of fish, no less—offer clues about the fundamental structure of our bodies.

How can we visualize events that happened millions and, in many cases, billions of years ago? Unfortunately, there were no eyewitnesses; none of us was around. In fact, nothing that talks or has a mouth or even a head was around for most of this time. Even worse, the animals that existed back then have been dead and buried for so long their bodies are only rarely preserved. If you consider that over 99 percent of all species that ever lived are now extinct, that only a very small fraction are preserved as fossils, and that an even smaller fraction still are ever found, then any attempt to see our past seems doomed from the start.

DIGGING FOSSILS—SEEING OURSELVES

I first saw one of our inner fish on a snowy July afternoon while studying 375-million-year-old rocks on Ellesmere Island, at a latitude about 80 degrees north. My colleagues and I had traveled up to this desolate part of the world to try to discover one of the key stages in the shift from fish to land-living animals. Sticking out of the rocks was the snout of a fish. And not just any fish: a fish with a flat head. Once we saw the flat head we knew we were on to something. If more of this skeleton were found inside the cliff, it would reveal the early stages in the history of our skull, our neck, even our limbs.

What did a flat head tell me about the shift from sea to land? More relevant to my personal safety and comfort, why was I in the Arctic and not in Hawaii? The answers to these questions lie in the story of how we find fossils and how we use them to decipher our own past.

Fossils are one of the major lines of evidence that we use to understand ourselves. (Genes and embryos are others, which I will discuss later.) Most people do not know that finding fossils is something we can often do with surprising precision and predictability. We work at home to maximize our chances of success in the field. Then we let luck take over.

The paradoxical relationship between planning and chance is best described by Dwight D. Eisenhower's famous remark about warfare: "In preparing for battle, I have found that planning is essential, but plans are useless." This captures field paleontology in a nutshell. We make all kinds of plans to get us to promising fossil sites. Once we're there, the entire field plan may be thrown out the window. Facts on the ground can change our best-laid plans.

Yet we can design expeditions to answer specific scientific questions. Using a few simple ideas, which I'll talk about below, we can predict where important fossils might be found. Of course, we are not successful 100 percent of the time, but we strike it rich often enough to make things interesting. I have made a career out of doing just that: finding early mammals to answer questions of mammal origins, the earliest frogs to answer questions of frog origins, and some of the earliest limbed animals to understand the origins of land-living animals.

In many ways, field paleontologists have a significantly easier time finding new sites today than we ever did before. We know more about the geology of local areas, thanks to the geological exploration undertaken by local governments and oil and gas companies. The Internet gives us rapid access to maps, survey information, and aerial photos. I can even scan your backyard for promising fossil sites right from my laptop. To top it off, imaging and radiographic devices can see through some kinds of rock and allow us to visualize the bones inside.

Despite these advances, the hunt for the important fossils is much what it was a hundred years ago. Paleontologists still need to look at rock—literally to crawl over it—and the fossils within must often be removed by hand. So many decisions need to be made when prospecting for and removing fossil bone that these processes are difficult to automate. Besides, looking at a monitor screen to find fossils would never be nearly as much fun as actually digging for them.

What makes this tricky is that fossil sites are rare. To maximize our odds of success, we look for the convergence of three things. We look for places that have rocks of the right age, rocks of the right type to preserve fossils, and rocks that are exposed at the surface. There is another factor: serendipity. That I will show by example.

Our example will show us one of the great transitions in the history of life: the invasion of land by fish. For billions of years, all life lived only in water. Then, as of about 365 million years ago, creatures also inhabited land. Life in these two environments is radically different. Breathing in water requires very different organs than breathing in air. The same is true for excretion, feeding, and moving about. A whole new kind of body had to arise. At first glance, the divide between the two environments appears almost unbridgeable. But everything changes when we look at the evidence; what looks impossible actually happened.

In seeking rocks of the right age, we have a remarkable fact on our side. The fossils in the rocks of the world are not arranged at random. Where they sit, and what lies inside them, is most definitely ordered, and we can use this order to design our expeditions. Billions of years of change have left layer upon layer of different kinds of rock in the earth. The working assumption, which is easy to test, is that rocks on the top are younger than rocks on the bottom; this is usually true in areas that have a straightforward, layer-cake arrangement (think the Grand Canyon). But movements of the earth's crust can cause faults that shift the position of the layers, putting older rocks on top of younger ones. Fortunately, once the positions of these faults are recognized, we can often piece the original sequence of layers back together.

The fossils inside these rock layers also follow a progression, with lower layers containing species entirely different from those in the layers above. If we could quarry a single column of rock that contained the entire history of life, we would find an extraordinary range of fossils. The lowest layers would contain little visible evidence of life. Layers above them would contain impressions of a diverse set of jellyfish-like things. Layers still higher would have creatures with skeletons, appendages, and various organs, such as eyes. Above those would be layers with the first animals to have backbones. And so on. The layers with the first people would be found higher still. Of course, a single column containing the entirety of earth history does not exist. Rather, the rocks in each location on earth represent only a small sliver of time. To get the whole picture, we need to put the pieces together by comparing the rocks themselves and the fossils inside them, much as if working a giant jigsaw puzzle.

That a column of rocks has a progression of fossil species probably comes as no surprise. Less obvious is that we can make detailed predictions about what the species in each layer might actually look like by comparing them with species of animals that are alive today; this information helps us to predict the kinds of fossils we will find in ancient rock layers. In fact, the fossil sequences in the world's rocks can be predicted by comparing ourselves with the animals at our local zoo or aquarium.

How can a walk through the zoo help us predict where we should look in the rocks to find important fossils? A zoo offers a great variety of creatures that are all distinct in many ways. But let's not focus on what makes them distinct; to pull off our prediction, we need to focus on what different creatures share. We can then use the features common to all species to identify groups of creatures with similar traits. All the living things can be organized and arranged like a set of Russian nesting dolls, with smaller groups of animals comprised in bigger groups of animals. When we do this, we discover something very fundamental about nature.

Every species in the zoo and the aquarium has a head and two eyes. Call these species "Everythings." A subset of the creatures with a head and two eyes has limbs. Call the limbed species "Everythings with limbs." A subset of these headed and limbed creatures has a huge brain, walks on two feet, and speaks. That subset is us, humans. We could, of course, use this way of categorizing things to make many more subsets, but even this threefold division has predictive power.

The fossils inside the rocks of the world generally follow this order, and we can put it to use in designing new expeditions. To use the example above, the first member of the group "Everythings," a creature with a head and two eyes, is found in the fossil record well before the first "Everything with limbs." More precisely, the first fish (a card-carrying member of the "Everythings") appears before the first amphibian (an "Everything with limbs"). Obviously, we refine this by looking at more kinds of animals and many more characteristics that groups of them share, as well as by assessing the actual age of the rocks themselves.

In our labs, we do exactly this type of analysis with thousands upon thousands of characteristics and species. We look at every bit of anatomy we can, and often at large chunks of DNA. There is so much data that we often need powerful computers to show us the groups within groups. This approach is the foundation of biology, because it enables us to make hypotheses about how creatures are related to one another.

Besides helping us refine the groupings of life, hundreds of years of fossil collection have produced a vast library, or catalogue, of the ages of the earth and the life on it. We can now identify general time periods when major changes occurred. Interested in the origin of mammals? Go to rocks from the period called the Early Mesozoic; geochemistry tells us that these rocks are likely about 210 million years old. Interested in the origin of primates? Go higher in the rock column, to the Cretaceous period, where rocks are about 80 million years old.

The order of fossils in the world's rocks is powerful evidence of our connections to the rest of life. If, digging in 600-million-year-old rocks, we found the earliest jellyfish lying next to the skeleton of a woodchuck, then we would have to rewrite our texts. That woodchuck would have appeared earlier in the fossil record than the first mammal, reptile, or even fish—before even the first worm. Moreover, our ancient woodchuck would tell us that much of what we think we know about the history of the earth and life on it is wrong. Despite more than 150 years of people looking for

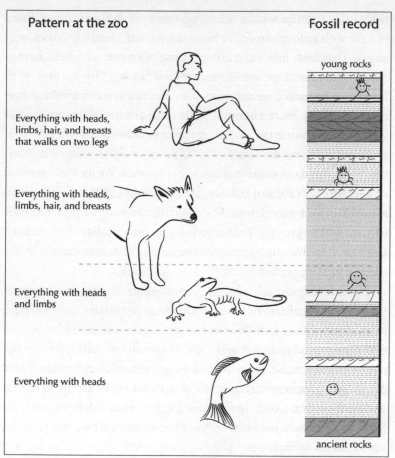

Pattern at the zoo

Everything with heads, limbs, hair, and breasts that walks on two legs

Everything with heads, limbs, hair, and breasts

Everything with heads and limbs

Everything with heads

Fossil record

young rocks

ancient rocks

What we discover on our walk through the zoo mirrors how fossils are laid out in the rocks of the world.

fossils—on every continent of earth and in virtually every rock layer that is accessible—this observation has never been made.

Let's now return to our problem of how to find relatives of the first fish to walk on land. In our grouping scheme, these creatures are somewhere between the "Everythings" and the "Everythings

with limbs." Map this to what we know of the rocks, and there
is strong geological evidence that the period from 380 million to
365 million years ago is the critical time. The younger rocks in that
range, those about 360 million years old, include diverse kinds of
fossilized animals that we would all recognize as amphibians or
reptiles. My colleague Jenny Clack at Cambridge University and
others have uncovered amphibians from rocks in Greenland that
are about 365 million years old. With their necks, their ears, and
their four legs, they do not look like fish. But in rocks that are
about 385 million years old, we find whole fish that look like, well,
fish. They have fins, conical heads, and scales; and they have no
necks. Given this, it is probably no great surprise that we should
focus on rocks about 375 million years old to find evidence of the
transition between fish and land-living animals.

We have settled on a time period to research, and so have identified the layers of the geological column we wish to investigate.
Now the challenge is to find rocks that were formed under conditions capable of preserving fossils. Rocks form in different kinds
of environments and these initial settings leave distinct signatures
on the rock layers. Volcanic rocks are mostly out. No fish that
we know of can live in lava. And even if such a fish existed, its
fossilized bones would not survive the superheated conditions in
which basalts, rhyolites, granites, and other igneous rocks are
formed. We can also ignore metamorphic rocks, such as schist and
marble, for they have undergone either superheating or extreme
pressure since their initial formation. Whatever fossils might have
been preserved in them have long since disappeared. Ideal to preserve fossils are sedimentary rocks: limestones, sandstones, silt-
stones, and shales. Compared with volcanic and metamorphic
rocks, these are formed by more gentle processes, including the
action of rivers, lakes, and seas. Not only are animals likely to live
in such environments, but the sedimentary processes make these
rocks more likely places to preserve fossils. For example, in an

ocean or lake, particles constantly settle out of the water and are deposited on the bottom. Over time, as these particles accumulate, they are compressed by new, overriding layers. The gradual compression, coupled with chemical processes happening inside the rocks over long periods of time, means that any skeletons contained in the rocks stand a decent chance of fossilizing. Similar processes happen in and along streams. The general rule is that the gentler the flow of the stream or river, the better preserved the fossils.

Every rock sitting on the ground has a story to tell: the story of what the world looked like as that particular rock formed. Inside the rock is evidence of past climates and surroundings often vastly different from those of today. Sometimes, the disconnect between present and past could not be sharper. Take the extreme example of Mount Everest, near whose top, at an altitude of over five miles, lie rocks from an ancient sea floor. Go to the North Face almost within sight of the famous Hillary Step, and you can find fossilized seashells. Similarly, where we work in the Arctic, temperatures can reach minus 40 degrees Fahrenheit in the winter. Yet inside some of the region's rocks are remnants of an ancient tropical delta, almost like the Amazon: fossilized plants and fish that could have thrived only in warm, humid locales. The presence of warm-adapted species at what today are extreme altitudes and latitudes attests to how much our planet can change: mountains rise and fall, climates warm and cool, and continents move about. Once we come to grips with the vastness of time and the extraordinary ways our planet has changed, we will be in a position to put this information to use in designing new fossil-hunting expeditions.

If we are interested in understanding the origin of limbed animals, we can now restrict our search to rocks that are roughly 375 million to 380 million years old and that were formed in oceans, lakes, or streams. Rule out volcanic rocks and metamorphic rocks, and our search image for promising sites comes into better focus.

We are only partly on the way to designing a new expedition, however. It does us no good if our promising sedimentary rocks of the right age are buried deep inside the earth, or if they are covered with grass, or shopping malls, or cities. We'd be digging blindly. As you can imagine, drilling a well hole to find a fossil offers a low probability of success, rather like throwing darts at a dartboard hidden behind a closet door.

The best places to look are those where we can walk for miles over the rock to discover areas where bones are "weathering out." Fossil bones are often harder than the surrounding rock and so erode at a slightly slower rate and present a raised profile on the rock surface. Consequently, we like to walk over bare bedrock, find a smattering of bones on the surface, then dig in.

So here is the trick to designing a new fossil expedition: find rocks that are of the right age, of the right type (sedimentary), and well exposed, and we are in business. Ideal fossil-hunting sites have little soil cover and little vegetation, and have been subject to few human disturbances. Is it any surprise that a significant fraction of discoveries happen in desert areas? In the Gobi Desert. In the Sahara. In Utah. In Arctic deserts, such as Greenland.

This all sounds very logical, but let's not forget serendipity. In fact, it was serendipity that put our team onto the trail of our inner fish. Our first important discoveries didn't happen in a desert, but along a roadside in central Pennsylvania where the exposures could hardly have been worse. To top it off, we were looking there only because we did not have much money.

It takes a lot of money and time to go to Greenland or the Sahara Desert. In contrast, a local project doesn't require big research grants, only money for gas and turnpike tolls. These are critical variables for a young graduate student or a newly hired college teacher. When I started my first job in Philadelphia, the lure was a group of rocks collectively known as the Catskill Formation of Pennsylvania. This formation has been extensively studied for over 150 years. Its age was well known and spanned the Late Devonian. In addition, its rocks were perfect to preserve early limbed animals and their closest relatives. To understand this, it is best to have an image of what Pennsylvania looked like back in the Devonian. Remove the image of present-day Philadelphia, Pittsburgh, or Harrisburg from your mind and think of the Amazon River delta. There were highlands in the eastern part of the state. A series of streams running east to west drained these mountains, ending in a large sea where Pittsburgh is today.

It is hard to imagine better conditions to find fossils, except that central Pennsylvania is covered in towns, forests, and fields. As for the exposures, they are mostly where the Pennsylvania Department of Transportation (PennDOT) has decided to put big roads. When PennDOT builds a highway, it blasts. When it blasts, it exposes rock. It's not always the best exposure, but we take what we can get. With cheap science, you get what you pay for.

And then there is also serendipity of a different order: in 1993, Ted Daeschler arrived to study paleontology under my supervision. This partnership was to change both our lives. Our different temperaments are perfectly matched: I have ants in my pants and am always thinking of the next place to look; Ted is patient and knows when to sit on a site to mine it for its riches. Ted and I began a survey of the Devonian rocks of Pennsylvania in hopes of finding new evidence on the origin of limbs. We began by driving to virtually every large roadcut in the eastern part of the state. To our great surprise, shortly after we began the survey, Ted found a marvelous shoulder bone. We named its owner *Hynerpeton*, a name that translates from Greek as "little creeping animal from Hyner." Hyner, Pennsylvania, is the nearest town. *Hynerpeton* had a very robust shoulder, which indicates a creature that likely had very powerful appendages. Unfortunately, we were never able to find the whole skeleton of the animal. The exposures were too

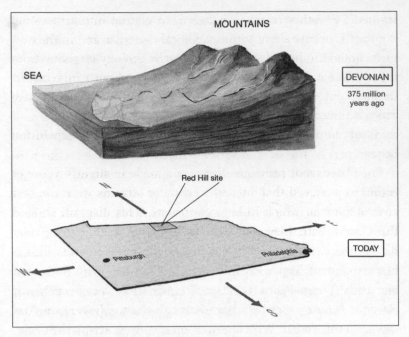

Along the roads in Pennsylvania, we were looking at an ancient river delta, much like the Amazon today. The state of Pennsylvania (bottom) with the Devonian topography mapped above it.

limited. By? You guessed it: vegetation, houses, and shopping malls.

After the discovery of *Hynerpeton* and other fossils from these rocks, Ted and I were champing at the bit for better-exposed rock. If our entire scientific enterprise was going to be based on recovering bits and pieces, then we could address only very limited questions. So we took a "textbook" approach, looking for well-exposed rocks of the right age and the right type in desert regions, meaning that we wouldn't have made the biggest discovery of our careers if not for an introductory geology textbook.

Originally we were looking at Alaska and the Yukon as potential venues for a new expedition, largely because of relevant discoveries made by other teams. We ended up getting into a bit of an argument/debate about some geological esoterica, and in the heat of the moment, one of us pulled the lucky geology textbook from a desk. While riffling through the pages to find out which one of us was right, we found a diagram. The diagram took our breath away; it showed everything we were looking for.

The argument stopped, and planning for a new field expedition began.

On the basis of previous discoveries made in slightly younger rocks, we believed that ancient freshwater streams were the best environment in which to begin our hunt. This diagram showed three areas with Devonian freshwater rocks, each with a river delta system. First, there is the east coast of Greenland. This is home to Jenny Clack's fossil, a very early creature with limbs and one of the earliest known tetrapods. Then there is eastern North America, where we had already worked, home to *Hynerpeton*. And there is a third area, large and running east–west across the Canadian Arctic. There are no trees, dirt, or cities in the Arctic. The chances were good that rocks of the right age and type would be extremely well exposed.

The Canadian Arctic exposures were well known, particularly to the Canadian geologists and paleobotanists who had already mapped them. In fact, Ashton Embry, the leader of the teams that did much of this work, had described the geology of the Devonian Canadian rocks as identical in many ways to the geology of Pennsylvania's. Ted and I were ready to pack our bags the minute we read this phrase. The lessons we had learned on the highways of Pennsylvania could help us in the High Arctic of Canada.

Remarkably, the Arctic rocks are even older than the fossil beds of Greenland and Pennsylvania. So the area perfectly fit all three of our criteria: age, type, and exposure. Even better, it was unknown to vertebrate paleontologists, and therefore un-prospected for fossils.

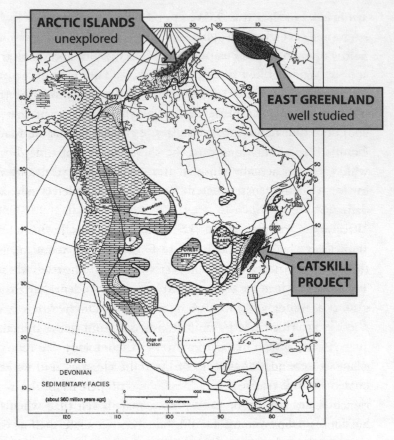

The map that started it all. This map of North America captures what we look for in a nutshell. The different kinds of shading reflect where Devonian age rocks, whether marine or freshwater, are exposed. Three areas that were once river deltas are labeled. Modified from figure 13.1, R. H. Dott and R. L. Batten, *Evolution of the Earth* (New York: McGraw-Hill, 1988). Reproduced with the permission of The McGraw-Hill Companies.

Our new challenges were totally different from those we faced in Pennsylvania. Along the highways in Pennsylvania, we risked being hit by the trucks that whizzed by as we looked for fossils. In the Arctic we risked being eaten by polar bears, running out of food, or being marooned by bad weather. No longer could we pack sandwiches in the car and drive to the fossil beds. We now had to spend at least eight days planning for every single day spent in the field, because the rocks were accessible only by air and the nearest supply base was 250 miles away. We could fly in only enough food and supplies for our crew, plus a slender safety margin. And, most important, the plane's strict weight limits meant that we could take out only a small fraction of the fossils that we found. Couple those limitations with the short window of time during which we can actually work in the Arctic every year, and you can see that the frustrations we faced were completely new and daunting.

Enter my graduate adviser, Dr. Farish A. Jenkins, Jr., from Harvard. Farish had led expeditions to Greenland for years and had the experience necessary to pull this venture off. The team was set. Three academic generations: Ted, my former student; Farish, my graduate adviser; and I were going to march up to the Arctic to try to discover evidence of the shift from fish to land-living animal.

There is no field manual for Arctic paleontology. We received gear recommendations from friends and colleagues, and we read books—only to realize that nothing could prepare us for the experience itself. At no time is this more sharply felt than when the helicopter drops one off for the first time in some godforsaken part of the Arctic totally alone. The first thought is of polar bears. I can't tell you how many times I've scanned the landscape looking for white specks that move. This anxiety can make you see things. In our first week in the Arctic, one of the crew saw a moving white speck. It looked like a polar bear about a quarter mile away. We scrambled like Keystone Kops for our guns, flares, and whistles until we discovered that our bear was a white Arctic hare two hundred feet away. With no trees or houses by which to judge distance, you lose perspective in the Arctic.

The Arctic is a big, empty place. The rocks we were interested in are exposed over an area about 1,500 kilometers wide. The creatures we were looking for were about four feet long. Somehow, we needed to home in on a small patch of rock that had preserved our fossils. Reviewers of grant proposals can be a ferocious lot; they light on this kind of difficulty all the time. A reviewer for one of Farish's early Arctic grant proposals put it best. As this referee wrote in his review of the proposal (not cordially, I might add), the odds of finding new fossils in the Arctic were "worse than finding the proverbial needle in the haystack."

It took us four expeditions to Ellesmere Island over six years to find our needle. So much for serendipity.

We found what we were looking for by trying, failing, and learning from our failures. Our first sites, in the 1999 field season, were way out in the western part of the Arctic, on Melville Island. We did not know it, but we had been dropped off on the edge of an ancient ocean. The rocks were loaded with fossils, and we found many different kinds of fish. The problem was that they all seemed to be deep-water creatures, not the kind we would expect to find in the shallow streams or lakes that gave rise to land-living animals. Using Ashton Embry's geological analysis, in 2000 we decided to move the expedition east to Ellesmere Island, because there the rocks would contain ancient streambeds. It did not take long for us to begin finding pieces of fish bones about the size of a quarter preserved as fossils.

The real breakthrough came toward the end of the field season in 2000. It was just before dinner, about a week before our scheduled pickup to return home. The crew had come back to camp, and we were involved in our early-evening activities: organizing the day's collections, preparing field notes, and beginning to assemble dinner. Jason Downs, then a college undergraduate eager to learn paleontology, hadn't returned to camp on time. This is a cause for worry, as we typically go out in teams; or if we separate, we give each other a definite schedule of when we will make contact again. With polar bears in the area and fierce storms that can roll in

Our camp (top) looks tiny in the vastness of the landscape. My summer home (bottom) is a small tent, usually surrounded by piles of rocks to protect it from fifty-mile-per-hour winds. Photographs by the author.

unexpectedly, we do not take any chances. I remember sitting in the main tent with the crew, the worry about Jason building with each passing moment. As we began to concoct a search plan, I heard the zipper on the tent open. At first all I saw was Jason's head. He had a wild-eyed expression on his face and was out of breath. As Jason entered the tent, we knew we were not dealing with a polar bear emergency; his shotgun was still shouldered. The cause of his delay became clear as his still shaking hand pulled out handful after handful of fossil bones that had been stuffed into every pocket: his coat, pants, inner shirt, and daypack. I imagine he would have stuffed his socks and shoes if he could have walked home that way. All of these little fossil bones were on the surface of a small site, no bigger than a parking spot for a compact car, about a mile away from camp. Dinner could wait.

With twenty-four hours of daylight in the Arctic summer, we did not have to worry about the setting sun, so we grabbed chocolate bars and set off for Jason's site. It was on the side of a hill between two beautiful river valleys and, as Jason had discovered, was covered in a carpet of fossil fish bones. We spent a few hours picking up the fragments, taking photos, and making plans. This site had all the makings of precisely what we were looking for. We returned the next day with a new goal: to find the exact layer of rock that contained the bones.

The trick was to identify the source of Jason's mess of bone fragments—our only hope of finding intact skeletons. The problem was the Arctic environment. Each winter, the temperature sinks to minus 40 degrees Fahrenheit. In the summer, when the sun never sets, the temperature rises to nearly 50 degrees. The resulting freeze-thaw cycle crumbles the surface rocks and fossils. Each winter they cool and shrink; each summer they heat and expand. As they shrink and swell with each season over thousands of years at the surface, the bones fall apart. Confronted by a jumbled mass of bone spread across the hill, we could not identify any obvious rock layer as their source. We spent several days following the fragment trails, digging test pits, practically using our geological hammers as divining rods to see where in the cliff the bones were emerging. After four days, we exposed the layer and eventually found skeleton upon skeleton of fossil fish, often lying one on

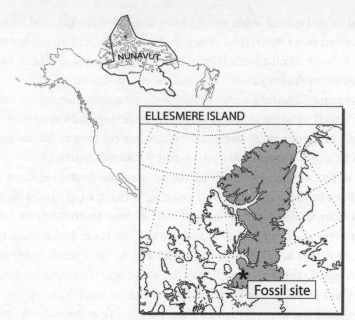

This is where we work: southern Ellesmere Island, in Nunavut Territory, Canada, 1,000 miles from the North Pole.

top of another. We spent parts of two summers exposing these fish.

Failure again: all the fish we were finding were well-known species that had been collected in sites of a similar age in Eastern Europe. To top it off, these fish weren't very closely related to land-living animals. In 2004, we decided to give it one more try. This was a do-or-die situation. The Arctic expeditions were prohibitively expensive and, short of a remarkable discovery, we would have to call it quits.

Everything changed over a period of four days in early July 2004. I was flipping rock at the bottom of the quarry, cracking ice more often than rock. I cracked the ice and saw something that I will never forget: a patch of scales unlike anything else we had yet seen in the quarry. This patch led to another blob covered by ice. It looked like a set of jaws. They were, however, unlike the jaws of any fish I had ever seen. They looked as if they might have connected to a flat head.

One day later, my colleague Steve Gatesy was flipping rocks at the top of the quarry. Steve removed a fist-size rock to reveal the snout of an animal looking right out at him. Like my ice-covered fish at the bottom of the pit, it had a flat head. It was new and important. But unlike my fish, Steve's had real potential. We were looking at the front end, and with luck the rest of the skeleton might be safely sitting in the cliff. Steve spent the rest of the summer removing rock from it bit by bit so that we could bring the entire skeleton back to the lab and clean it up. Steve's masterful work with this specimen led to the recovery of one of the finest fossils discovered to date at the water–land transition.

The specimens we brought back to the lab at home were little more than boulders with fossils inside. Over the course of two months, the rock was removed piece by piece, often manually with dental tools or small picks by the preparators in the lab. Every day a new piece of the fossil creature's anatomy was revealed. Almost every time a large section was exposed, we learned something new about the origin of land-living animals.

What we saw gradually emerge from these rocks during the fall of 2004 was a beautiful intermediate between fish and land-living animals. Fish and land-living animals differ in many respects. Fish have conical heads, whereas the earliest land-living animals have almost crocodile-like heads—flat, with the eyes on top. Fish do not have necks: their shoulders are attached to their heads by a series of bony plates. Early land-living animals, like all their descendants, do have necks, meaning their heads can bend independently of their shoulders.

There are other big differences. Fish have scales all over their bodies; land-living animals do not. Also, importantly, fish have

The process of finding fossils begins with a mass in a rock that is gradually removed over time. Here I show a fossil as it travels from the field to the lab and is carefully prepared as a specimen: the skeleton of the new animal. Photograph in upper left by author; other photographs courtesy of Ted Daeschler, Academy of Natural Sciences of Philadelphia.

fins, whereas land-living animals have limbs with fingers, toes, wrists, and ankles. We can continue these comparisons and make a very long list of the ways that fish differ from land-living animals.

But our new creature broke down the distinction between these two different kinds of animal. Like a fish, it has scales on its back and fins with fin webbing. But, like early land-living animals, it has a flat head and a neck. And, when we look inside the fin, we see bones that correspond to the upper arm, the forearm, even parts of the wrist. The joints are there, too: this is a fish with shoulder, elbow, and wrist joints. All inside a fin with webbing.

Virtually all of the features that this creature shares with land-living creatures look very primitive. For example, the shape and various ridges on the fish's upper "arm" bone, the humerus, look part fish and part amphibian. The same is true of the shape of the skull and the shoulder.

It took us six years to find it, but this fossil confirmed a prediction of paleontology: not only was the new fish an intermediate between two different kinds of animal, but we had found it also *in the right time period in earth's history* and *in the right ancient environment*. The answer came from 375-million-year-old rocks, formed in ancient streams.

As the discoverers of the creature, Ted, Farish, and I had the privilege of giving it a formal scientific name. We wanted the name to reflect the fish's provenance in the Nunavut Territory of the Arctic and the debt we owed to the Inuit people for permission to work there. We engaged the Nunavut Council of Elders, for-

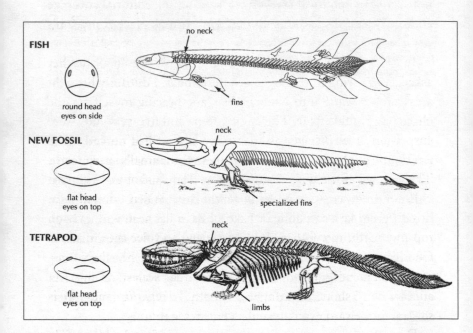

This figure says it all. *Tiktaalik* is intermediate between fish and primitive land-living animal.

mally known as the Inuit Qaujimajatuqangit Katimajiit, to come up with a name in the Inuktitut language. My obvious concern was that a committee named Inuit Qaujimajatuqangit Katimajiit might not propose a scientific name we could pronounce. I sent them a picture of the fossil, and the elders came up with two suggestions, *Siksagiaq* and *Tiktaalik*. We went with *Tiktaalik* for its relative ease of pronunciation for the non-Inuktitut-speaking tongue and because of its meaning in Inuktitut: "large freshwater fish."

Tiktaalik was the lead story in a number of newspapers the day after the find was announced in April 2006, including above-the-fold headlines in such places as *The New York Times*. This attention ushered in a week unlike any other in my normally quiet life. Though for me the greatest moment of the whole media blitz was not seeing the political cartoons or reading the editorial coverage and the heated discussions on the blogs. It took place at my son's preschool.

In the midst of the press hubbub, my son's preschool teacher asked me to bring in the fossil and describe it. I dutifully brought a cast of *Tiktaalik* into Nathaniel's class, bracing myself for the chaos that would ensue. The twenty four- and five-year-olds were surprisingly well behaved as I described how we had worked in the Arctic to find the fossil and showed them the animal's sharp teeth. Then I asked what they thought it was. Hands shot up. The first child said it was a crocodile or an alligator. When queried why, he said that like a crocodile or lizard it has a flat head with eyes on top. Big teeth, too. Other children started to voice their dissent. Choosing the raised hand of one of these kids, I heard: No, no, it isn't a crocodile, it is a fish, because it has scales and fins. Yet another child shouted, "Maybe it is both." *Tiktaalik*'s message is so straightforward even preschoolers can see it.

For our purposes, there is an even more profound take on *Tik-taalik*. This fish doesn't just tell us about fish; it also contains a piece of us. The search for this connection is what led me to the Arctic in the first place.

How can I be so sure that this fossil says something about my own body? Consider the neck of *Tiktaalik*. All fish prior to *Tiktaalik* have a set of bones that attach the skull to the shoulder, so that every time the animal bent its body, it also bent its head. *Tiktaalik* is different. The head is completely free of the shoulder. This whole arrangement is shared with amphibians, reptiles, birds, and mammals, including us. The entire shift can be traced to the loss of a few small bones in a fish like *Tiktaalik*.

I can do a similar analysis for the wrists, ribs, ears, and other parts of our skeleton—all these features can be traced back to a fish like this. This fossil is just as much a part of our history as the

Tracing arm bones from fish to humans.

African hominids, such as *Australopithecus afarensis*, the famous "Lucy." Seeing Lucy, we can understand our history as highly advanced primates. Seeing *Tiktaalik* is seeing our history as fish.

So what have we learned? Our world is so highly ordered that we can use a walk through a zoo to predict the kinds of fossils that lie in the different layers of rocks around the world. Those predictions can bring about fossil discoveries that tell us about ancient events in the history of life. The record of those events remains inside us, as part of our anatomical organization.

What I haven't mentioned is that we can also trace our history inside our genes, through DNA. This record of our past doesn't lie in the rocks of the world; it lies in every cell inside us. We'll use both fossils and genes to tell our story, the story of the making of our bodies.

CHAPTER TWO

GETTING A GRIP

I mages of the medical school anatomy lab are impossible to forget. Imagine walking into a room where you will spend several
months taking a human body apart layer by layer, organ by organ,
all as a way to learn tens of thousands of new names and body
structures.

In the months before I did my first human dissection, I readied
myself by trying to envision what I would see, how I would react,
and what I would feel. It turned out that my imagined world in
no way prepared me for the experience. The moment when we
removed the sheet and saw the body for the first time wasn't
nearly as stressful as I'd expected. We were to dissect the chest, so
we exposed it while leaving the head, arms, and legs wrapped
in preservative-drenched gauze. The tissues did not look very
human. Having been treated with a number of preservatives, the
body didn't bleed when cut, and the skin and internal organs had
the consistency of rubber. I began to think that the cadaver looked
more like a doll than a human. A few weeks went by as we exposed
the organs of the chest and abdomen. I came to think that I was
quite the pro; having already seen most of the internal organs, I
had developed a cocky self-confidence about the whole experience. I did my initial dissections, made my cuts, and learned the
anatomy of most of the major organs. It was all very mechanical,
detached, and scientific.

This comfortable illusion was rudely shattered when I uncovered the hand. As I unwrapped the gauze from the fingers—as I saw the joints, fingertips, and fingernails for the first time—I uncovered emotions that had been concealed during the previous few weeks. This was no doll or mannequin; this had once been a living person, who used that hand to carry and caress. Suddenly, this mechanical exercise, dissection, became deeply and emotionally personal. Until that moment, I was blind to my connection to the cadaver. I had already exposed the stomach, the gallbladder, and other organs; but what sane person forms a human connection at the sight of a gallbladder?

What is it about a hand that seems quintessentially human? The answer must, at some level, be that the hand is a visible connection between us; it is a signature for who we are and what we can attain. Our ability to grasp, to build, and to make our thoughts real lies inside this complex of bones, nerves, and vessels.

The immediate thing that strikes you when you see the inside of the hand is its compactness. The ball of your thumb, the thenar eminence, contains four different muscles. Twiddle your thumb and tilt your hand: ten different muscles and at least six different bones work in unison. Inside the wrist are at least eight small bones that move against one another. Bend your wrist, and you are using a number of muscles that begin in your forearm, extending into tendons as they travel down your arm to end at your hand. Even the simplest motion involves a complex interplay among many parts packed in a small space.

The relationship between complexity and humanity within our hands has long fascinated scientists. In 1822, the eminent Scottish surgeon Sir Charles Bell wrote the classic book on the anatomy of hands. The title says it all: *The Hand, Its Mechanism and Vital Endowments as Evincing Design*. To Bell, the structure of the hand was "perfect" because it was complex and ideally arranged for the way we live. In his eye, this designed perfection could only have a divine origin.

The great anatomist Sir Richard Owen was one of the scientific leaders in this search for divine order within bodies. He was fortunate to be an anatomist in the mid-1800s, when there were still entirely new kinds of animals to discover living in the distant reaches of the earth. As more and more parts of the world were explored by westerners, all sorts of exotic creatures made their way back to laboratories and museums. Owen described the first gorilla, brought back from expeditions to central Africa. He coined the name "dinosaur" for a new kind of fossil creature discovered in rocks in England. His study of these bizarre new creatures gave him special insights: he began to see important patterns in the seeming chaos of life's diversity.

Owen discovered that our arms and legs, our hands and feet, fit into a larger scheme. He saw what anatomists before him had long known, that there is a pattern to the skeleton of a human arm: one bone in the upper arm, two bones in the forearm, a bunch of nine little bones at the wrists, then a series of five rods that make the fingers. The pattern of bones in the human leg is much the same: one bone, two bones, lotsa blobs, and five toes. In comparing this pattern with the diversity of skeletons in the world, Owen made a remarkable discovery.

Owen's genius was not that he focused on what made the various skeletons different. What he found, and later promoted in a series of lectures and volumes, were *exceptional similarities* among creatures as different as frogs and people. All creatures with limbs, whether those limbs are wings, flippers, or hands, have a common design. One bone, the humerus in the arm or the femur in the leg, articulates with two bones, which attach to a series of small blobs, which connect with the fingers or toes. This pattern underlies the architecture of all limbs. Want to make a bat wing? Make the fingers really long. Make a horse? Elongate the middle fingers and toes and reduce and lose the outer ones. How about a frog leg? Elongate the bones of the leg and fuse several of them together.

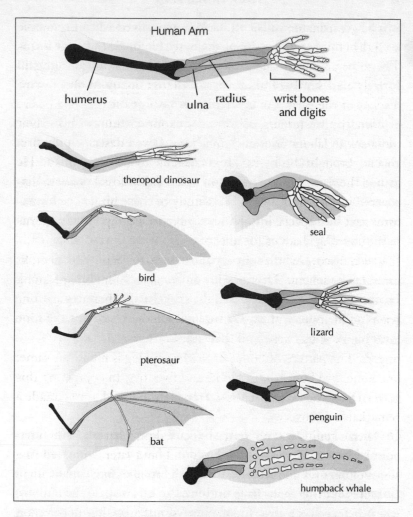

The common plan for all limbs: one bone, followed by two bones, then little blobs, then fingers or toes.

The differences between creatures lie in differences in the shapes and sizes of the bones and the numbers of blobs, fingers, and toes. Despite radical changes in what limbs do and what they look like, this underlying blueprint is always present.

For Owen, seeing a design in the limbs was only the beginning:

when he looked at skulls and backbones, indeed when he considered the entire architecture of the body, he found the same thing. There is a fundamental design in the skeleton of all animals. Frogs, bats, humans, and lizards are all just variations on a theme. That theme, to Owen, was the plan of the Creator.

Shortly after Owen announced this observation in his classic monograph *On the Nature of Limbs,* Charles Darwin supplied an elegant explanation for it. The reason the wing of a bat and the arm of a human share a common skeletal pattern is because they shared a common ancestor. The same reasoning applies to human arms and bird wings, human legs and frog legs—everything that has limbs. There is a major difference between Owen's theory and that of Darwin: Darwin's theory allows us to make very precise predictions. Following Darwin, we would expect to find that Owen's blueprint has a history that will be revealed in creatures with no limbs at all. Where, then, do we look for the history of the limb pattern? We look to fish and their fin skeletons.

SEEING THE FISH

In Owen and Darwin's day, the gulf between fins and limbs seemed impossibly wide. Fish fins don't have any obvious similarities to limbs. On the outside, most fish fins are largely made up of fin webbing. Our limbs have nothing like this, nor do the limbs of any other creature alive today. The comparisons do not get any easier when you remove the fin webbing to see the skeleton inside. In most fish, there is nothing that can be compared to Owen's one bone–two bones–lotsa blobs–digits pattern. All limbs have a single long bone at their base: the humerus in the upper arm and the femur in the upper leg. In fish, the whole skeleton looks utterly different. The base of a typical fin has four or more bones inside.

In the mid-1800s, anatomists began to learn of mysterious living fish from the southern continents. One of the first was discovered by German anatomists working in South America. It looked like a normal fish, with fins and scales, but behind its throat were large vascular sacs: lungs. Yet the creature had scales and fins. So confused were the discoverers that they named the creature *Lepidosiren paradoxa*, "paradoxically scaled amphibian." Other fish with lungs, aptly named lungfish, were soon found in Africa and Australia. African explorers brought one to Owen. Scientists such as Thomas Huxley and the anatomist Carl Gegenbaur found lungfish to be essentially a cross between an amphibian and a fish. Locals found them delicious.

A seemingly trivial pattern in the fins of these fish had a profound impact on science. The fins of lungfish have at their base a single bone that attaches to the shoulder. To anatomists, the comparison was obvious. Our upper arm has a single bone, and that single bone, the humerus, attaches to our shoulder. In the lungfish, we have a fish with a humerus. And, curiously, it is not just any fish; it is a fish that also has lungs. Coincidence?

As a handful of these living species became known in the 1800s, clues started to come from another source. As you might guess, these insights came from ancient fish.

One of the first of these fossils came from the shores of the Gaspé Peninsula in Quebec, in rocks about 380 million years old. The fish was given a tongue-twister name, *Eusthenopteron*. *Eusthenopteron* had a surprising mix of features seen in amphibians and fish. Of Owen's one bone–two bones–lotsa blobs– digits plan of limbs, *Eusthenopteron* had the one bone–two bones part, but in a fin. Some fish, then, had structures like those in a limb. Owen's archetype was not a divine and eternal part of all life. It had a history, and that history was to be found in Devonian age rocks, rocks that are between 390 million and 360 million years old. This profound insight defined a whole new research program with a whole new research agenda: somewhere in the Devonian rocks we should find the origin of fingers and toes.

In the 1920s, the rocks provided more surprises. A young Swedish paleontologist, Gunnar Säve-Söderbergh, was given the extraordinary opportunity to explore the east coast of Greenland for fossils. The region was terra incognita, but Säve-Söderbergh recognized that it featured enormous deposits of Devonian rocks. He was one of the exceptional field paleontologists of all time, who throughout his short career uncovered remarkable fossils with both a bold exploring spirit and a precise attention to detail. (Unfortunately, he was to die tragically of tuberculosis at a young age, soon after the stunning success of his field expeditions.) In expeditions between 1929 and 1934, Säve-Söderbergh's team discovered what, at the time, was labeled a major missing link. Newspapers around the world trumpeted his discovery; editorials analyzed its importance; cartoons lampooned it. The fossils in question were true mosaics: they had fish-like heads and tails, yet they also had fully formed limbs (with fingers and toes), and vertebrae that were extraordinarily amphibian-like. After Säve-Söderbergh died, the fossils were described by his colleague Erik Jarvik, who named one of the new species *Ichthyostega soderberghi* in honor of his friend.

For our story, *Ichthyostega* is a bit of a letdown. True, it is a remarkable intermediate in most aspects of its head and back, but it says very little about the origin of limbs because, like any amphibian, it already has fingers and toes. Another creature, one that received little notice when Säve-Söderbergh announced it, was to provide real insights decades later. This second limbed animal was to remain an enigma until 1988, when a paleontological colleague of mine, Jenny Clack, who we introduced in the first chapter, returned to Säve-Söderbergh's sites and found more of its fossils. The creature, called *Acanthostega gunnari* back in the 1920s on the basis of Säve-Söderbergh's fragments, now revealed

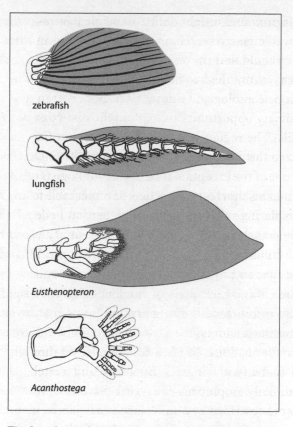

zebrafish

lungfish

Eusthenopteron

Acanthostega

The fins of most fish—for example, a zebrafish (top)—have large amounts of fin webbing and many bones at the base. Lungfish captured people's interest because like us they have a single bone at the base of the appendage. *Eusthenopteron* (middle) showed how fossils begin to fill the gap; it has bones that compare to our upper arm and forearm. *Acanthostega* (bottom) shares *Eusthenopteron*'s pattern of arm bones with the addition of fully formed digits.

full limbs, with fingers and toes. But it also carried a real surprise: Jenny found that the limb was shaped like a flipper, almost like that of a seal. This suggested to her that the earliest limbs arose to help animals swim, not walk. That insight was a significant advance, but a problem remained: *Acanthostega* had fully formed digits, with a real wrist and no fin webbing. *Acanthostega* had a limb, albeit a very primitive one. The search for the origins of hands and feet, wrists and ankles had to go still deeper in time. This is where matters stood until 1995.

FINDING FISH FINGERS AND WRISTS

In 1995, Ted Daeschler and I had just returned to his house in Philadelphia after driving all through central Pennsylvania in an effort to find new roadcuts. We had found a lovely cut on Route 15 north of Williamsport, where PennDOT had created a giant cliff in sandstones about 365 million years old. The agency had dynamited the cliff and left piles of boulders alongside the highway. This was perfect fossil-hunting ground for us, and we stopped to crawl over the boulders, many of them roughly the size of a small microwave oven. Some had fish scales scattered throughout, so we decided to bring a few back home to Philadelphia. Upon our return to Ted's house, his four-year-old daughter, Daisy, came running out to see her dad and asked what we had found.

In showing Daisy one of the boulders, we suddenly realized that sticking out of it was a sliver of fin belonging to a large fish. We had completely missed it in the field. And, as we were to learn, this was no ordinary fish fin: it clearly had lots of bones inside. People in the lab spent about a month removing the fin from the boulder—and there, exposed for the first time, was a fish with Owen's pattern. Closest to the body was one bone. This one bone attached to two bones. Extending away from the fin were about eight rods. This looked for all the world like a fish with fingers.

Our fin had a full set of webbing, scales, and even a fish-like shoulder, but deep inside were bones that corresponded to much of the "standard" limb. Unfortunately, we had only an isolated fin.

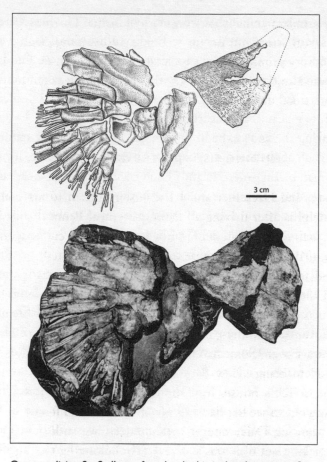

3 cm

Our tantalizing fin. Sadly, we found only this isolated specimen. Stipple diagram used with the permission of Scott Rawlins, Arcadia University. Photo by the author.

What we needed was to find a place where whole bodies of creatures could be recovered intact. A single isolated fin could never help us answer the real questions: What did the creature use its fins for, and did the fish fins have bones and joints that worked like ours? The answer would come only from whole skeletons.

For that find, we had to search almost ten years. And I wasn't

the first to recognize what we were looking at. The first were two
professional fossil preparators, Fred Mullison and Bob Masek.
Preparators use dental tools to scratch at the rocks we find in the
field and thereby expose the fossils inside. It can take months, if
not years, for a preparator to turn a big fossil-filled boulder like
ours into a beautiful, research-quality specimen.

During the 2004 expedition, we had collected three chunks of
rock, each about the size of a piece of carry-on luggage, from the
Devonian of Ellesmere Island. Each contained a flat-headed animal: the one I found in ice at the bottom of the quarry, Steve's
specimen, and a third specimen we discovered in the final week of
the expedition. In the field we had removed each head, leaving
enough rock intact around it to explore in the lab for the rest of
the body. Then the rocks were wrapped in plaster for the trip
home. Opening these kinds of plaster coverings in the lab is much
like encountering a time capsule. Bits and pieces of our life on the
Arctic tundra are there, as are the field notes and scribbles we
make on the specimen. Even the smell of the tundra comes waft-
ing out of these packages as we crack the plaster open.

Fred in Philadelphia and Bob in Chicago were scratching on
different boulders at the same general time. From one of these
Arctic blocks, Bob had pulled out a particular small bone in a big
fin of the Fish (we hadn't named it *Tiktaalik* yet). What made this
cube-shaped blob of bone different from any other fin bone was a
joint at the end that had spaces for four other bones. That is, the
blob looked scarily like a wrist bone—but the fins in the block that
Bob was preparing were too jumbled to tell for sure. The next
piece of evidence came from Philadelphia a week later. Fred, a
magician with his dental tools, uncovered a whole fin in his block.
At the right place, just at the end of the forearm bones, the fin
had *that* bone. And *that* bone attached to four more beyond. We
were staring at the origin of a piece of our own bodies inside this
375-million-year-old fish. We had a fish with a wrist.

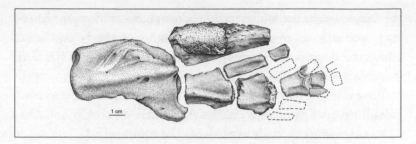

The bones of the front fin of *Tiktaalik*—a fish with a wrist.

Over the next months, we were able to see much of the rest of the appendage. It was part fin, part limb. Our fish had fin webbing, but inside was a primitive version of Owen's one bone–two bones–lotsa blobs–digits arrangement. Just as Darwin's theory predicted: at the right time, at the right place, we had found intermediates between two apparently different kinds of animals.

Finding the fin was only the beginning of the discovery. The real fun for Ted, Farish, and me came from understanding what the fin did and how it worked, and in guessing why a wrist joint arose in the first place. Solutions to these puzzles are found in the structure of the bones and joints themselves.

When we took the fin of *Tiktaalik* apart, we found something truly remarkable: all the joint surfaces were extremely well preserved. *Tiktaalik* has a shoulder, elbow, and wrist composed of the same bones as an upper arm, forearm, and wrist in a human. When we study the structure of these joints to assess how one bone moves against another, we see that *Tiktaalik* was specialized for a rather extraordinary function: it was capable of doing push-ups.

When we do push-ups, our hands lie flush against the ground, our elbows are bent, and we use our chest muscles to move up and down. *Tiktaalik*'s body was capable of all of this. The elbow was capable of bending like ours, and the wrist was able to bend to make the fish's "palm" lie flat against the ground. As for chest muscles, *Tiktaalik* likely had them in abundance. When we look at the shoulder and the underside of the arm bone at the point where they would have connected, we find massive crests and scars where the large pectoral muscles would have attached. *Tiktaalik* was able to "drop and give us twenty."

Why would a fish ever want to do a push-up? It helps to consider the rest of the animal. With a flat head, eyes on top, and ribs, *Tiktaalik* was likely built to navigate the bottom and shallows of streams or ponds, and even to flop around on the mudflats along the banks. Fins capable of supporting the body would have been very helpful indeed for a fish that needed to maneuver in all these environments. This interpretation also fits with the geology of the site where we found the fossils of *Tiktaalik*. The structure of the rock layers and the pattern of the grains in the rocks themselves have the characteristic signature of a deposit that was origi-

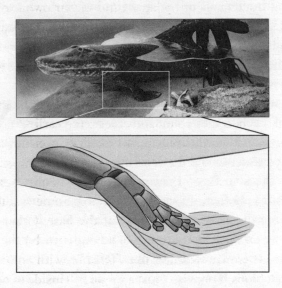

A full-scale model of *Tiktaalik*'s body (top) and a drawing of its fin (bottom). This is a fin in which the shoulder, elbow, and proto-wrist were capable of performing a type of push-up.

nally formed by a shallow stream surrounded by large seasonal mudflats.

But why live in these environments at all? What possessed fish to get out of the water or live in the margins? Think of this: virtually every fish swimming in these 375-million-year-old streams was a predator of some kind. Some were up to sixteen feet long, almost twice the size of the largest *Tiktaalik*. The most common fish species we find alongside *Tiktaalik* is seven feet long and has a head as wide as a basketball. The teeth are barbs the size of railroad spikes. Would you want to swim in these ancient streams?

It is no exaggeration to say that this was a fish-eat-fish world. The strategies to succeed in this setting were pretty obvious: get big, get armor, or get out of the water. It looks as if our distant ancestors avoided the fight.

But this conflict avoidance meant something much deeper to us. We can trace many of the structures of our own limbs to the fins of these fish. Bend your wrist back and forth. Open and close your hand. When you do this, you are using joints that first appeared in the fins of fish like *Tiktaalik*. Earlier, these joints did not exist. Later, we find them in limbs.

Proceed from *Tiktaalik* to amphibians all the way to mammals, and one thing becomes abundantly clear: the earliest creature to have the bones of our upper arm, our forearm, even our wrist and palm, also had scales and fin webbing. That creature was a fish.

What do we make of the one bone–two bones–lotsa blobs– digits plan that Owen attributed to a Creator? Some fish, for example the lungfish, have the one bone at the base. Other fish, for example *Eusthenopteron*, have the one bone–two bones arrangement. Then there are creatures like *Tiktaalik*, with one bone–two bones–lotsa blobs. There isn't just a single fish inside of our limbs; there is a whole aquarium. Owen's blueprint was assembled in fish.

Tiktaalik might be able to do a push-up, but it could never throw a baseball, play the piano, or walk on two legs. It is a long

way from *Tiktaalik* to humanity. The important, and often surprising, fact is that most of the major bones humans use to walk,
throw, or grasp first appear in animals tens to hundreds of millions of years before. The first bits of our upper arm and leg are in
380-million-year-old fish like *Eusthenopteron*. *Tiktaalik* reveals the
early stages in the evolution of our wrist, palm, and finger area.
The first true fingers and toes are seen in 365-million-year-old
amphibians like *Acanthostega*. Finally, the full complement of wrist
and ankle bones found in a human hand or foot is seen in reptiles
more than 250 million years old. The basic skeleton of our hands
and feet emerged over hundreds of millions of years, first in fish
and later in amphibians and reptiles.

But what are the major changes that enable us to use our hands
or walk on two legs? How do these shifts come about? Let's look at
two simple examples from limbs for some answers.

We humans, like many other mammals, can rotate our thumb
relative to our elbow. This simple function is very important for the
use of our hands in everyday life. Imagine trying to eat, write, or
throw a ball without being able to rotate your hand relative to your
elbow. We can do this because one forearm bone, the radius, rotates
along a pivot point at the elbow joint. The structure of the joint at
the elbow is wonderfully designed for this function. At the end of
our upper-arm bone, the humerus, lies a ball. The tip of the radius,
which attaches here, forms a beautiful little socket that fits on the
ball. This ball-and-socket joint allows the rotation of our hand,
called pronation and supination. Where do we see the beginnings
of this ability? In creatures like *Tiktaalik*. In *Tiktaalik,* the end of
the humerus forms an elongated bump onto which a cup-shaped
joint on the radius fits. When *Tiktaalik* bent its elbow, the end of its
radius would rotate, or pronate, relative to the elbow. Refinements
of this ability are seen in amphibians and reptiles, where the end of
the humerus becomes a true ball, much like our own.

Looking now at the hind limb, we find a key feature that gives us the capacity to walk, one we share with other mammals. Unlike fish and amphibians, our knees and elbows face in opposite directions. This feature is critical: think of trying to walk with your kneecap facing backward. A very different situation exists in fish like *Eusthenopteron*, where the equivalents of the knee and elbow face largely in the same direction. We start development with little limbs oriented much like those in *Eusthenopteron*, with elbows and knees facing in the same direction. As we grow in the womb, our knees and elbows rotate to give us the state of affairs we see in humans today.

Our bipedal pattern of walking uses the movements of our hips, knees, ankles, and foot bones to propel us forward in an upright stance unlike the sprawled posture of creatures like *Tiktaalik*. One big difference is the position of our hips. Our legs do not project sideways like those of a crocodile, amphibian, or fish; rather, they project underneath our bodies. This change in posture came about by changes to the hip joint, pelvis, and upper leg: our pelvis became bowl shaped, our hip socket became deep, our femur gained its distinctive neck, the feature that enables it to project under the body rather than to the side.

Do the facts of our ancient history mean that humans are not special or unique among living creatures? Of course not. In fact, knowing something about the deep origins of humanity only adds to the remarkable fact of our existence: all of our extraordinary capabilities arose from basic components that evolved in ancient fish and other creatures. From common parts came a very unique construction. We are not separate from the rest of the living world; we are part of it down to our bones and, as we will see shortly, even our genes.

In retrospect, the moment when I first saw the wrist of a fish was as meaningful as the first time I unwrapped the fingers of the cadaver back in the human anatomy lab. Both times I was uncovering a deep connection between my humanity and another being.

CHAPTER THREE

HANDY GENES

While my colleagues and I were digging up the first *Tiktaalik* in the Arctic in July 2004, Randy Dahn, a researcher in my laboratory, was sweating it out on the South Side of Chicago doing genetic experiments on the embryos of sharks and skates, cousins of stingrays. You've probably seen small black egg cases, known as mermaid's purses, on the beach. Inside the purse once lay an egg with yolk, which developed into an embryonic skate or ray. Over the years, Randy has spent hundreds of hours experimenting with the embryos inside these egg cases, often working well past midnight. During the fateful summer of 2004, Randy was taking these cases and injecting a molecular version of vitamin A into the eggs. After that he would let the eggs develop for several months until they hatched.

His experiments may seem to be a bizarre way to spend the better part of a year, let alone for a young scientist to launch a promising scientific career. Why sharks? Why a form of vitamin A?

To make sense of these experiments, we need to step back and look at what we hope they might explain. What we are really getting at in this chapter is the recipe, written in our DNA, that builds our bodies from a single egg. When sperm fertilizes an egg, that fertilized egg does not contain a tiny hand, for instance. The hand is built from the information contained in that single cell. This takes us to a very profound problem. It is one thing to compare the bones of our hands with the bones in fish fins. What happens if you compare the genetic recipe that builds our hands with the recipe that builds a fish's fin? To find answers to this question, just like Randy, we will follow a trail of discovery that takes us from our hands to the fins of sharks and even to the wings of flies.

As we've seen, when we discover creatures that reveal different and often simpler versions of our bodies inside their own, a wonderfully direct window opens into the distant past. But there is a big limitation to working with fossils. We cannot do experiments on long-dead animals. Experiments are great because we can actually manipulate something to see the results. For this reason, my laboratory is split directly in two: half is devoted to fossils, the other half to embryos and DNA. Life in my lab can be schizophrenic. The locked cabinet that holds *Tiktaalik* specimens is adjacent to the freezer containing our precious DNA samples.

Experiments with DNA have enormous potential to reveal inner fish. What if you could do an experiment in which you treated the embryo of a fish with various chemicals and actually changed its body, making part of its fin look like a hand? What if you could show that the genes that build a fish's fin are virtually the same as those that build our hands?

We begin with an apparent puzzle. Our body is made up of hundreds of different kinds of cells. This cellular diversity gives our tissues and organs their distinct shapes and functions. The cells that make our bones, nerves, guts, and so on look and behave entirely differently. Despite these differences, there is a deep similarity among every cell inside our bodies: all of them contain exactly the same DNA. If DNA contains the information to build our bodies, tissues, and organs, how is it that cells as different as those found in muscle, nerve, and bone contain the same DNA?

The answer lies in understanding what pieces of DNA (the genes) are actually turned on in every cell. A skin cell is different from a neuron because different genes are active in each cell. When a gene is turned on, it makes a protein that can affect what the cell looks like and how it behaves. Therefore, to understand what makes a cell in the eye different from a cell in the bones of the hand, we need to know about the genetic switches that control the activity of genes in each cell and tissue.

Here's the important fact: these genetic switches help to assemble us. At conception, we start as a single cell that contains all the DNA needed to build our body. The plan for that entire body unfolds via the instructions contained in this single micro-scopic cell. To go from this generalized egg cell to a complete human, with trillions of specialized cells organized in just the right way, whole batteries of genes need to be turned on and off at just the right stages of development. Like a concerto composed of individual notes played by many instruments, our bodies are a composition of individual genes turning on and off inside each cell during our development.

This information is a boon to those who work to understand bodies, because we can now compare the activity of different genes to assess what kinds of changes are involved in the origin of new organs. Take limbs, for example. When we compare the ensemble of genes active in the development of a fish fin to those

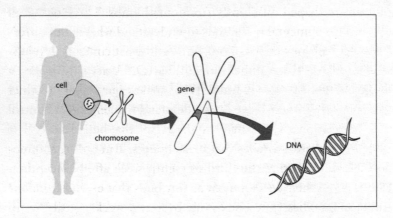

Genes are stretches of DNA contained in every cell of our bodies.

active in the development of a human hand, we can catalogue the genetic differences between fins and limbs. This kind of comparison gives us some likely culprits—the genetic switches that may have changed during the origin of limbs. We can then study what these genes are doing in the embryo and how they might have changed. We can even do experiments in which we manipulate the genes to see how bodies actually change in response to different conditions or stimuli.

To see the genes that build our hands and feet, we need to take a page from a script for the TV show *CSI: Crime Scene Investigation*—start at the body and work our way in. We will begin by looking at the structure of our limbs, and zoom all the way down to the tissues, cells, and genes that make it.

MAKING HANDS

Our limbs exist in three dimensions: they have a top and a bottom, a pinky side and a thumb side, a base and a tip. The bones at the tips, in our fingers, are different from the bones at the shoulder. Likewise, our hands are different from one side to the other. Our pinkies are shaped differently from our thumbs. The Holy Grail of our developmental research is to understand what genes differentiate the various bones of our limb, and what controls development in these three dimensions. What DNA actually makes a pinky different from a thumb? What makes our fingers distinct from our arm bones? If we can understand the genes that control such patterns, we will be privy to the recipe that builds us.

All the genetic switches that make fingers, arm bones, and toes do their thing during the third to eighth week after conception. Limbs begin their development as tiny buds that extend from our embryonic bodies. The buds grow over two weeks, until the tip forms a little paddle. Inside this paddle are millions of cells which

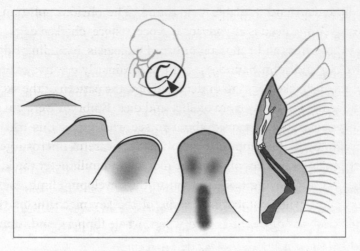

The development of a limb, in this case a chicken wing. All of the key
stages in the development of a wing skeleton happen inside the egg.

will ultimately give rise to the skeleton, nerves, and muscles that
we'll have for the rest of our lives.

To study how this pattern emerges, we need to look at embryos
and sometimes interfere with their development to assess what
happens when things go wrong. Moreover, we need to look at
mutants and at their internal structures and genes, often by making whole mutant populations through careful breeding. Obviously, we cannot study humans in these ways. The challenge for
the pioneers in this field was to find the animals that could be
useful windows into our own development. The first experimental
embryologists interested in limbs in the 1930s and 1940s faced
several problems. They needed an organism in which the limbs
were accessible for observation and experiment. The embryo had
to be relatively large, so that they could perform surgical proce-
dures on it. Importantly, the embryo had to grow in a protected
place, in a container that sheltered it from jostling and other environmental disturbances. Also, and critically, the embryos had to

be abundant and available year-round. The obvious solution to this scientific need is at your local grocery store: chicken eggs.

In the 1950s and 1960s a number of biologists, including Edgar Zwilling and John Saunders, did extraordinarily creative experiments on chicken eggs to understand how the pattern of the skeleton forms. This was an era of slice and dice. Embryos were cut up and various tissues moved about to see what effect this had on development. The approach involved very careful microsurgery, manipulating patches of tissue no more than a millimeter thick. In that way, by moving tissues about in the developing limb, Saunders and Zwilling uncovered some of the key mechanisms that build limbs as different as bird wings, whale flippers, and human hands.

They discovered that two little patches of tissue essentially control the development of the pattern of bones inside limbs. A strip of tissue at the extreme end of the limb bud is essential for *all* limb development. Remove it, and development stops. Remove it early, and we are left with only an upper arm, or a piece of an arm. Remove it slightly later, and we end up with an upper arm and a forearm. Remove it even later, and the arm is almost complete, except that the digits are short and deformed.

Another experiment, initially done by Mary Gasseling in John Saunders's laboratory, led to a powerful new line of research. Take a little patch of tissue from what will become the pinky side of a limb bud, early in development, and transplant it on the opposite side, just under where the first finger will form. Let the chick develop and form a wing. The result surprised nearly everybody. The wing developed normally except that it also had a *full duplicate set of digits*. Even more remarkable was the pattern of the digits: the new fingers were mirror images of the normal set. Obviously, something inside that patch of tissue, some molecule or gene, was able to direct the development of the pattern of the fingers. This result spawned a blizzard of new experiments, and we learned that this effect can be mimicked by a variety of other means. For example, take a chicken embryo and dab a little vitamin A on its limb bud, or simply inject vitamin A into the egg, and let the embryo develop. If you supply the vitamin A at the right concentration and at the right stage, you'll get the same mirror-image duplication that Gasseling, Saunders, and Zwilling got from the grafting experiments. This patch of tissue was named the zone of polarizing activity (ZPA). Essentially, the ZPA is a patch of tissue that causes the pinky side to be different from the thumb side. Obviously chicks do not have a pinky and a thumb. The terminology we use is to number the digits, with our pinky corresponding to digit five of other animals and our thumb corresponding to digit one.

The ZPA drew interest because it appeared, in some way, to control the formation of fingers and toes. But how? Some people believed that the cells in the ZPA made a molecule that then spread across the limb to instruct cells to make different fingers.

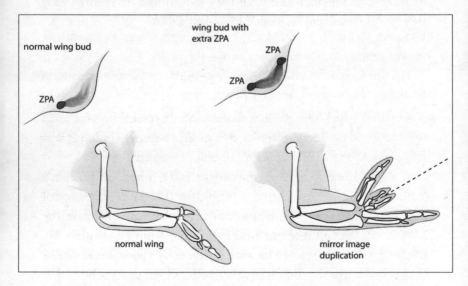

Moving a little patch of tissue called the ZPA causes the fingers to be duplicated.

The key proposal was that it was the concentration of this un-named molecule that was the important factor. In areas close to the ZPA, where there is a high concentration of this molecule, cells would respond by making a pinky. In the opposite side of the developing hand, farther from the ZPA so that the molecule was more diffused, the cells would respond by making a thumb. Cells in the middle would each respond according to the concentration of this molecule to make the second, third, and fourth fingers.

This concentration-dependent idea could be tested. In 1979, Denis Summerbell placed an extremely small piece of foil be-tween the ZPA patch and the rest of the limb. The idea was to use this barrier to prevent any kind of molecule from diffusing from the ZPA to the other side. Summerbell studied what happened to the cells on each side of the barrier. Cells on the ZPA side formed digits. Cells on the opposite side often did not form digits; if they did, the digits were badly malformed. The conclusion was obvious. Something was emanating from the ZPA that controlled how the digits formed and what they looked like. To identify that something, researchers needed to look at DNA.

THE DNA RECIPE

That project was left to a new generation of scientists. Not until the 1990s, when new molecular techniques became available, was the genetic control for the ZPA's operation unraveled.

A major breakthrough happened in 1993, when Cliff Tabin's laboratory at Harvard started hunting for the genes that control the ZPA. Their prey was the molecular mechanisms that gave the ZPA its ability to make our pinky different from our thumb. By the time his group started to work in the early 1990s, a number of experiments like the ones I've described had led us to believe that some sort of molecule caused the whole thing. This was a grand theory, but nobody knew what this molecule was. People would propose one molecule after another, only to find that none was up to the job. Finally, the Tabin lab came up with a novel notion, and one very relevant to the theme of this book. Look to flies for the answer.

Genetic experiments in the 1980s had revealed the wonderful pattern of gene activity that sculpts the body of a fly from a single-celled egg. The body of a fruit fly is organized from front to back, with the head at the front and the wings at the back. Whole batteries of genes are turned on and off during fly development, and this pattern of gene activity serves to demarcate the different regions of the fly.

Tabin didn't know it at the time, but two other laboratories— those of Andy MacMahon and Phil Ingham—had already come up with the same general idea independently. What emerged was a remarkably successful collaboration among three different lab groups. One of the fly genes caught the attention of Tabin, McMahon, and Ingham. They noted that this gene made one end of a body segment look different from the other. Fly geneticists named it *hedgehog*. Doesn't the function of *hedgehog* in the fly body—to make one region different from another—sound like what the ZPA does in making the pinky different from the thumb? That parallel was not lost on the three labs. So off they went, looking for a *hedgehog* gene in creatures like chickens, mice, and fish.

Because the lab groups knew the structure of the fly's *hedgehog* gene, they had a search image to help them single out the gene in chickens. Each gene has a distinctive sequence; using a number of molecular tools, the researchers could scan the chicken's DNA for the *hedgehog* sequence. After a lot of trial and error, they found a chicken *hedgehog* gene.

Just as paleontologists get to name new species, geneticists get to name new genes. The fly geneticists who discovered *hedgehog* had named it that because the flies with a mutation in the gene had bristles that reminded them of a little hedgehog. Tabin, McMahon, and Ingham named the chicken version of the gene *Sonic hedgehog,* after the Sega Genesis video game.

Now came the fun question: What does *Sonic hedgehog* actually do in the limb? The Tabin group attached a dye to a molecule that would stick to the gene, enabling them to visualize where the gene is active in the limb. To their great surprise, they found that only cells in a tiny patch of the limb had gene activity: the ZPA.

So the next steps became obvious. The patterns of activity in the *Sonic hedgehog* gene should mimic those of the ZPA tissue itself. Recall that when you treat the limb with retinoic acid, a form of vitamin A, you get a ZPA active on the opposite side. Guess what happens when you treat a limb with retinoic acid, then map where *Sonic hedgehog* is active? *Sonic hedgehog* becomes active on both sides—pinky and thumb—just as the ZPA does when it is treated with retinoic acid.

Knowing the structure of the chicken *Sonic hedgehog* gave other researchers the tools to look for it in everything else that has fingers, from frogs to humans. Every limbed animal has the *Sonic hedgehog* gene. And in every single animal that we have studied, *Sonic hedgehog* is active in the ZPA tissue. If *Sonic hedgehog* hadn't turned on properly during the eighth week of your own development, then you either would have extra fingers or your pinky and thumb would look alike. Occasionally, when things go wrong with *Sonic hedgehog,* the hand ends up looking like a broad paddle with as many as twelve fingers that all look alike.

We now know that *Sonic hedgehog* is one of dozens of genes that act to sculpt our limbs from shoulder to fingertip by turning on and off at the right time. Remarkably, work in chickens, frogs, and mice was telling us the same thing. The DNA recipe to build upper arms, forearms, wrists, and digits is virtually identical in every creature that has limbs.

How far back can we trace *Sonic hedgehog* and the other bits of

DNA that build limbs? Is this stuff active in building the skeleton of fish fins? Or are hands genetically completely different from fish fins? We saw an inner fish in the anatomy of our arms and hands. What about the DNA that builds it?

Enter Randy Dahn with his mermaid's purses.

GIVING SHARKS A HAND

Randy Dahn entered my laboratory with a simple but very elegant idea: treat skate embryos just the way Cliff Tabin treated chicken eggs. Randy's goal was to perform all the experiments on skates that chicken biologists had performed on chicken eggs, from Saunders and Zwilling's tissue surgeries all the way to Cliff Tabin's gene experiments. Skates develop in an egg with a kind of shell and a yolk. Skates even have big embryos, just as chickens do. Because of these convenient facts, we could apply to skates many of the genetic and experimental tools people had developed to understand chickens.

What could we learn by comparing the development of a shark fin to that of a chicken leg? Even more relevant, what could we learn about ourselves from all this?

Chickens, as Saunders, Zwilling, and Tabin showed, are a surprisingly good proxy for our own limbs. Everything that was discovered by Saunders and Zwilling's cutting and grafting experiments and by Tabin's DNA work applies to our own limbs as well: we have a ZPA, we have *Sonic hedgehog*, and both have a great bearing on our well-being. As we saw, a malfunctioning ZPA or a mutation in *Sonic hedgehog* can cause major malformations in human hands.

Randy wanted to determine how different the apparatus is that builds our hands. How deep is our connection to the rest of life? Is the recipe that builds our hands new, or does it, too, have deep roots in other creatures? If so, how deep?

Sharks and their relatives are the earliest creatures that have fins with a skeleton inside. Ideally, to answer Randy's question, you would want to bring a 400-million-year-old shark fossil into the laboratory, grind it up, and look at its genetic structure. Then you'd try to manipulate its fossil embryos to learn whether *Sonic hedgehog* is active in the same general place as in our limbs today. This would be a wonderful experiment, but it is impossible. We cannot extract DNA from fossils so old, and, even if we could, we could never find embryos of those fossil animals on which to do experiments.

Living sharks and their relatives are the next best thing. Nobody would ever confuse a shark fin with a human hand: you couldn't ask for two more different kinds of appendages. Not only are sharks and humans very distantly related, but also the skeletal structures of their appendages look nothing alike. Nothing even remotely similar to Owen's one bone–two bones–lotsa blobs–digits pattern is inside a shark's fin. Instead, the bones inside are shaped like rods, long and short, thin and wide. We call them bones even though they are made of cartilage (sharks and skates are known as cartilaginous fish, because their skeletons never turn into hard bone). If you want to assess whether *Sonic hedgehog*'s role in limbs is unique to limbed animals, why not choose a species utterly different in almost every way? In addition, why not choose the species that is the most primitive living fish with any kind of paired appendage, whether fin or limb? Sharks fit both bills perfectly.

Our first problem was a simple one. We needed a reliable source for the embryos of sharks and skates. Sharks proved difficult to obtain with any degree of regularity, but skates, their close relatives, were another matter. So we started with sharks and used skates as our supply of sharks dwindled. We found a supplier who would ship us every month or two a batch of twenty or thirty egg cases with embryos inside. We became a virtual cargo cult as we waited each month for our shipment of precious egg cases.

Work by Tabin's group and others gave Randy important clues to begin his search. Since Tabin's work in 1993, people had found *Sonic hedgehog* in a number of different species, everything from fish to humans. With the knowledge of the structure of the gene, Randy was able to search all the DNA of the skate and shark for *Sonic hedgehog*. In a very short time he found it: a shark *Sonic hedgehog* gene.

The key questions to answer were Where is *Sonic hedgehog* active?, and, even more important, What is it doing?

The egg cases were put to use as Randy visualized where and when *Sonic hedgehog* is active in the development of skates. He first studied whether *Sonic hedgehog* turns on at the same time in skate fin development as it does in chicken limbs. Yes, it does. Then he studied whether it is turned on in the patch of tissue at the back end of the fin, the equivalent of our pinky. Yes again. Now he did his vitamin A experiment. This was the million-dollar moment. If you treat the limb of a chicken or mammal with this compound, you get a patch of tissue that has *Sonic hedgehog* activity on the opposite side, and this result is coupled with a duplication of the bones. Randy injected the egg, waited a day or so, and then checked whether, as in chickens, the vitamin A caused *Sonic hedgehog* to turn on in the opposite side of the limb. It did. Now came the long wait. We knew that *Sonic hedgehog* was behaving the same way in our hands and in skates' and sharks' fins. But what would the effect of all this be on the skeleton? We would have to wait two months for the answer.

The embryos were developing inside an opaque egg case. All we could tell was whether the creature was alive; the inside of the fin was invisible to us.

The end result was a stunning example of similarity among us, sharks, and skates: a mirror-image fin. The dorsal fins duplicated their structures in a wonderful front-to-back pattern, the same kind we saw with experiments in limbs. Limbs duplicate a limb structure. Shark fins duplicate a shark fin structure as do skates. *Sonic hedgehog* has a similar effect in even the most different kinds of appendage skeletons found on earth today.

One effect of *Sonic hedgehog*, you may recall, is to make the fingers distinct from one another. As we saw with respect to the ZPA, what kind of digit develops depends on how close the digit is to the source of *Sonic hedgehog*. A normal adult skate fin contains many skeletal rods, which all look alike. Could we make these rods different from one another, like our digits? Randy took a small bead impregnated with the protein made by *Sonic hedgehog* and put it in between these identical skeletal rods. The key to his experiment is that he used mouse *Sonic hedgehog*. So now we have a real contraption: a skate embryo with a bead inside that is gradually leaking mouse *Sonic hedgehog* protein. Would that mouse protein have any effect on a shark or a skate?

There are two extreme outcomes to an experiment like this. One is that nothing happens. This would mean that skates are so different from mice that *Sonic hedgehog* protein has no effect. The other extreme outcome would present a stunning example of our inner fish. This outcome would be that the rods develop differently from one another, demonstrating that *Sonic hedgehog* does something similar in skates and in us. And let's not forget that since Randy is using the protein from a mammal, it means that the genetic recipe would be really, really similar.

Not only did the rods end up looking different from one another, they responded to *Sonic hedgehog*, much as fingers do, on the basis of how close they were to the *Sonic hedgehog* bead: the closer rods developed a different shape from the ones farther

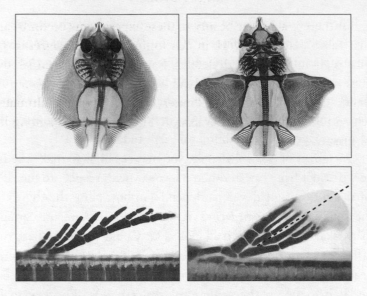

Normal fins (left) and Randy's treated fins. The treated fins showed a mirror-image duplication just as chicken wings did. Photographs courtesy of Randall Dahn, University of Chicago.

away. To top matters off, it was the mouse protein that did the job so effectively in the skates.

The "inner fish" that Randy found was not a single bone, or even a section of the skeleton. Randy's inner fish lay in the biological tools that actually build fins. Experiment after experiment on creatures as different as mice, sharks, and flies shows us that the lessons of *Sonic hedgehog* are very general. All appendages, whether they are fins or limbs, are built by similar kinds of genes. What does this mean for the problem we looked at in the first two chapters—the transition of fish fins into limbs? It means that this great evolutionary transformation did not involve the origin of new DNA: much of the shift likely involved using ancient genes, such as those involved in shark fin development, in new ways to make limbs with fingers and toes.

But there is a deeper beauty to these experiments on limbs and fins. Tabin's lab used work in *flies* to find a gene in *chickens* that tells us about *human* birth defects. Randy used the Tabin lab discovery to tell us something about our connections to *skates*. An "inner fly" helped find an "inner chicken," which ultimately helped Randy find an "inner skate." The connections among living creatures run deep.

CHAPTER FOUR

TEETH EVERYWHERE

The tooth gets short shrift in anatomy class: we spend all of five minutes on it. In the pantheon of favorite organs—I'll leave it to each of you to make your list—teeth rarely reach the top five. Yet the little tooth contains so much of our connection to the rest of life that it is virtually impossible to understand our bodies without knowing teeth. Teeth also have special significance for me, because it was in searching for them that I first learned how to find fossils and how to run a fossil expedition.

The job of teeth is to make bigger creatures into smaller pieces. When attached to a moving jaw, teeth slice, dice, and macerate. Mouths are only so big, and teeth enable creatures to eat things that are bigger than their mouths. This is particularly true of creatures that do not have hands or claws that can shred or cut things before they get to the mouth. True, big fish tend to eat littler fish. But teeth can be the great equalizer: smaller fish can munch on bigger fish if they have good teeth. Smaller fish can use their teeth to scrape scales, feed on particles, or take out whole chunks of flesh from bigger fish.

We can learn a lot about an animal by looking at its teeth. The bumps, pits, and ridges on teeth often reflect the diet. Carnivores, such as cats, have blade-like molars to cut meat, while plant eaters have a mouth full of flatter teeth that can macerate leaves and nuts. The informational value of teeth was not lost on the anatomists of history. The French anatomist Georges Cuvier once famously boasted that he could reconstruct an animal's entire skeleton from a single tooth. This is a little over the top, but the general point is valid; teeth are a powerful window into an animal's lifestyle.

Human mouths reveal that we are all-purpose eaters, for we have several kinds of teeth. Our front teeth, the incisors, are flat blades specialized for cutting. The rearmost teeth, the molars, are flatter, with a distinctive pattern that can macerate plant or animal tissue. The premolars, in between, are intermediate in function between incisors and molars.

The most remarkable thing about our mouths is the precision with which we chew. Open and close your mouth: your teeth always come together in the same position, with upper and lower teeth fitting together precisely. Because the upper and lower cusps, basins, and ridges match closely, we are able to break up food with maximal efficiency. In fact, a mismatch between upper and lower teeth can shatter our teeth, and enrich our dentists.

Paleontologists find teeth wonderfully informative. Teeth are the hardest parts of our bodies, because the enamel includes a high proportion of the mineral hydroxyapatite—higher even than is found in bones. Thanks to their hardness, teeth are often the best-preserved animal part we find in the fossil record for many time periods. This is lucky; since teeth are such a great clue to an animal's diet, the fossil record can give us a good window on how different ways of feeding came about. This is particularly true of mammal history: whereas many reptiles have similar teeth, those of mammals are distinctive. The mammal section of a typical pale-ontology course feels almost like Dentistry 101.

Living reptiles—crocodiles, lizards, snakes—lack much of what makes mammalian mouths unique. A crocodile's teeth, for example, all have a similar blade-like shape; the only difference between them is that some are big and others small. Reptiles also lack the precise occlusion—the fit between upper and lower teeth—that humans and other mammals have. Also, whereas we mammals replace our teeth only once, reptiles typically receive visits from the tooth fairy for their entire lives, replacing their teeth continually as they wear and break down.

A very basic piece of us—our mammalian way of precise chewing—emerges in the fossil record from around the world that ranges from 225 million to 195 million years ago. At the base, in the older rocks, we find a number of reptiles that look superficially dog-like. Walking on four legs, they have big skulls, and many of them have sharp teeth. There the resemblance stops. Unlike dogs, these reptiles have a jaw made up of many bones, and their teeth don't really fit well together. Also, their teeth are replaced in a decidedly reptilian way: new teeth pop in and out throughout the animals' lives.

Go higher in the rocks and we see something utterly different: the appearance of mammalness. The bones of the jaw get smaller and move to the ear. We can see the first evidence of upper and lower teeth coming together in precise ways. The jaw's shape changes, too: what was a simple rod in reptiles looks more like a boomerang in mammals. At this time, too, teeth are replaced only once per lifetime, as in us. We can trace all these changes in the fossil record, especially from certain sites in Europe, South Africa, and China.

The rocks of about 200 million years ago contain rodent-like creatures, such as *Morganucodon* and *Eozostrodon*, that have begun to look like mammals. These animals, no bigger than a mouse, hold important pieces of us inside. Pictures cannot convey just how wonderful these early mammals are. For me, it was a real thrill to see creatures like them for the first time.

When I entered graduate school, I wanted to study early mammals. I chose Harvard because Farish A. Jenkins, Jr., whom we met in the first chapter, was leading expeditions to the American West that systematically scoured the rocks for signs of how mammals developed their distinct abilities to chew. The work was real exploration; Farish and his team were looking for new localities and sites, not returning to places other people had discovered. Farish had assembled a talented group of fossil finders comprising staff from Harvard's Museum of Comparative Zoology and a few freelance mercenaries. Chief among them were Bill Amaral, Chuck Schaff, and the late Will Downs. These people were my introduction to the world of paleontology.

Farish and the team had studied geological maps and aerial photos to choose promising areas where they might find early mammals. Then, each summer, they got in their trucks and headed off into the deserts of Wyoming, Arizona, and Utah. By the time I joined them, in 1983, they had already found a number of important new mammals and fossil sites. I was struck by the power of predictions: simply by reading scientific articles and books, Farish's team could identify likely and unlikely places to find early mammals.

My baptism in field paleontology came from walking out in the Arizona desert with Chuck and Bill. At first, the whole enterprise seemed utterly random. I expected something akin to a military campaign, an organized and coordinated reconnaissance of the area. What I saw looked like the extreme opposite. The team would plunk down on a particular patch of rock, and people would scatter in every conceivable direction to look for fragments of bone on the surface. For the first few weeks of the expedition, they left me alone. I'd set off looking for fossils, systematically inspecting every rock I saw for a scrap of bone at the surface. At the end of each day we would come home to show off the goodies we found. Chuck would have several bags of bones. Bill would have his complement, usually with some sort of little skull or other prize. And I had nothing, my empty bag a sad reminder of how much I had to learn.

After a few weeks of this, I decided it would be a good idea to walk with Chuck. He seemed to have the fullest bags each day, so why not take some cues from the expert? Chuck was happy to walk with me and expound on his long career in field paleontology. Chuck is all West Texas with a Brooklyn flourish: cowboy boots and western values with a New York accent. While he regaled me with tales of his past expeditions, I found the whole experience utterly humbling. First, Chuck did not look at every rock, and when he chose one to look at, for the life of me I couldn't figure out why. Then there was the really embarrassing aspect of all this: Chuck and I would look at the same patch of ground. I saw nothing but rock—barren desert floor. Chuck saw fossil teeth, jaws, and even chunks of skull.

An aerial view would have shown two people walking alone in the middle of a seemingly limitless plain, where the vista of dusty red and green sandstone mesas, buttes, and badlands extended for miles. But Chuck and I were staring only at the ground, at the rubble and talus of the desert floor. The fossils we sought were tiny, no more than a few inches long, and ours was a very small world. This intimate environment stood in extreme contrast to the vastness of the desert panorama that surrounded us. I felt as if my walking partner was the only person on the entire planet, and my whole existence was focused on pieces of rubble.

Chuck was extraordinarily patient with me as I pestered him with questions for the better part of each day's walk. I wanted him to describe *exactly* how to find bones. Over and over, he told me to look for "something different," something that had the texture of bone not rock, something that glistened like teeth, something that looked like an arm bone, not a piece of sandstone. It sounded easy, but I couldn't grasp what he was telling me. Try as I might, I still returned home each day empty-handed. Now it was even more embarrassing, as Chuck, who was looking at the same rocks, came home with bag after bag.

Finally, one day, I saw my first piece of tooth glistening in the desert sun. It was sitting in some sandstone rubble, but there it was, as plain as day. The enamel had a sheen that no other rock had; it was like nothing I had seen before. Well, not exactly—I was looking at things like it every day. The difference was this time I finally saw it, saw the distinction between rock and bone. The tooth glistened, and when I saw it glisten I spotted its cusps. The whole isolated tooth was about the size of a dime, not including the roots that projected from its base. To me, it was as glorious as the biggest dinosaur in the halls of any museum.

All of a sudden, the desert floor exploded with bone; where once I had seen only rock, now I was seeing little bits and pieces of fossil everywhere, as if I were wearing a special new pair of glasses and a spotlight was shining on all the different pieces of bone. Next to the tooth were small fragments of other bones, then more teeth. I was looking at a jaw that had weathered out on the surface and fragmented. I started to return home with my own little bags each night.

Now that I could finally see bones for myself, what once seemed a haphazard group effort started to look decidedly ordered. People weren't just scattering randomly across the desert; there were real though unspoken rules. Rule number one: go to the most productive-looking rocks, judging by whatever search image or visual cues you've gained from previous experience. Rule number two: don't follow in anybody's footsteps; cover new ground (Chuck had graciously let me break this one). Rule three: if your plum area already has somebody on it, find a new plum, or search a less promising site. First come, first served.

Over time, I began to learn the visual cues for other kinds of bones: long bones, jawbones, and skull parts. Once you see these things you never lose the ability to find them. Just as a great fisherman can read the water and see the fish within, so a fossil finder uses a catalogue of search images that make fossils seem to jump out from the rocks. I was beginning to gain my own visual impressions of what fossil bones look like in different rocks and in different lighting conditions. Finding fossils in the morning sun is very different from finding them in the afternoon, because of the way the light plays along the ground.

Twenty years later, I know that I must go through a similar experience every time I look for fossils someplace new, from the Triassic of Morocco to the Devonian of Ellesmere Island. I'll struggle for the first few days, almost as I did those days with Chuck in Arizona twenty years ago. The difference is that now I have some confidence that a search image will kick in eventually.

The whole goal of the prospecting I did with Chuck was to find a site with enough bones to mark a fossil-rich layer that we could expose. By the time I joined the crew, Farish's team had already discovered such a zone, a patch of rock about a hundred feet long that contained skeleton after skeleton of small animals.

Farish's fossil quarry was in some very fine-grained mudstone. The trick to working on it was to realize that the fossils were coming from one thin layer, no more than a millimeter thick. Once you exposed that surface, you had a very good chance of seeing bones. They were tiny, no more than an inch or two long, and black, so they looked almost like black smudges against the brownish rock. The little animals we found included frogs (some of the earliest), legless amphibians, lizards and other reptiles, and, importantly, some of the earliest mammals.

The key point is that the early mammals were small. Very small. Their teeth were not much more than 2 millimeters long. To spot them, you had to be very careful and, more often, very lucky. If the tooth was covered by a crumb of rock or even by a few grains of sand, you might never see it.

It was the sight of these early mammals that really hooked me. I'd expose the fossil layer, then scan the entire surface through my 10-power hand lens. I'd scrutinize the whole thing on my hands and knees, with my eye and hand lens only about two inches from the surface of the ground. Thus engrossed, I'd often forget where I was and accidentally trespass on my neighbor's spot only to have a bag of dirt dumped on my head as a sharp reminder to keep to my space. Occasionally, though, I'd hit the jackpot and see a deep connection for the first time. The teeth would look like little blades, with cusps and roots. The cusps on those little teeth revealed something very special. Each tooth had a characteristic pattern of wear at the face where upper and lower teeth fit together. I was seeing some of the first evidence of our pattern of precise chewing, only in a tiny mammal 190 million years old.

The power of those moments was something I'll never forget. Here, cracking rocks in the dirt, I was discovering objects that could change the way people think. That juxtaposition between the most child-like, even humbling, activities and one of the great human intellectual aspirations has never been lost on me. I try to remind myself of it each time I dig somewhere new.

Returning to school that fall, I developed the expedition bug big-time. I wanted to lead my own expedition but lacked the resources to do anything big, so I set off to explore rocks in Connecticut that were about 200 million years old. Well studied during the nineteenth century, they had been the setting for a number of important fossil discoveries. I figured that if I hit those same rocks with my hand lens and my wonderfully successful early mammal search image, I'd find lots of goodies. I rented a minivan, grabbed a case of collecting bags, and set off.

Yet another lesson learned: I found nothing. Back to the drawing board, or more precisely, the geology library at school.

I needed a place where 200-million-year-old rocks were well exposed: in Connecticut there were only roadcuts. The ideal place would be along the coast, where wave action would provide lots of freshly broken rock surface to look at. Looking at a map made my choice clear: up in Nova Scotia, Triassic and Jurassic rocks (roughly 200 million years old) lay along the surface. To top it off, the tourist literature about the area advertised the world's highest tides, occasionally over fifty feet. I couldn't believe my luck.

I called the expert on these rocks, Paul Olsen, who had just started teaching at Columbia University. If I was excited about fossil-finding prospects before I talked to Paul, I was frothing afterward. He described the perfect geology for finding small mammals or reptiles: ancient streams and dunes that had just the right properties to preserve tiny bones. Even better, he had already found some dinosaur bones and footprints along a stretch of beach near the town of Parrsboro, Nova Scotia. Paul and I hatched a plan to visit Parrsboro together and scan the beach for little fossils. This was wonderfully generous on Paul's part because he had dibs on the area and was under no responsibility to help me out, let alone collaborate.

I consulted with Farish on my emerging plans, and he not only offered money but suggested that I take the fossil-finding experts, Bill and Chuck. Money, Bill, Chuck, Paul Olsen, excellent rocks, and decent exposures—what more could you want? The following summer, I led my very first fossil expedition.

Off I went in a rented station wagon to the beaches of Nova Scotia with my field crew, Bill and Chuck. The joke, of course, was on me. With Bill and Chuck along, who between them had more years of field experience than I had birthdays, I was the leader in name only. They called the fossil-finding shots, while I paid the dinner bills.

The rocks in Nova Scotia were exposed in absolutely gorgeous orange sandstone cliffs along the Bay of Fundy. The tides would go in and out about half a mile each day, exposing enormous flats of orange bedrock. It wasn't long before we started to find bones in many different areas. Small white flecks of bone were coming out along the cliffs. Paul was finding footprints everywhere, even in the flats opened by the moving tides each day.

Paul Olsen finding footprints in the tidal flats of Nova Scotia. At high tide, the water would come all the way to the cliffs at left. The arrowhead points to a spot where, if we timed our trip wrong, we would be stuck on the cliffs for hours at a time. Photograph by the author.

Chuck, Bill, Paul, and I spent two weeks digging in Nova Scotia, finding bits, flakes, and fragments of bones sticking out of the rocks. Bill, being the fossil preparator of the group, continually warned me not to expose much of the bones in the field but rather to wrap them up still covered in sandstone so that he could trace the bones in the laboratory under a microscope in more controlled conditions. We did this, but I'll admit to being disappointed with what we brought home: just a few shoeboxes of rocks, with small chips and flakes of bones showing. As we drove home, I recall thinking that even though we hadn't found much, it had been a great experience. Then I took a week's vacation; Chuck and Bill returned to the lab.

When I returned to Boston, Chuck and Bill were out to lunch. Some colleagues were visiting the museum and, having caught sight of me, came up to shake my hand, offer congratulations, and slap me on the back. I was being treated like a conquering hero, but I had no idea why; it seemed like a bizarre joke, as if they were setting me up for some big con. They told me to go to Bill's lab to see my trophy. Not knowing what to think, I ran.

Under Bill's microscope was a tiny jaw, not more than half an inch long. In it were a few minute teeth. The jaw's owner was clearly a reptile, because the teeth had only a single root at the base, whereas mammal teeth have many. But on the teeth were tiny bumps and ridges that I could see even with the naked eye. Looking at the teeth under the microscope gave me the biggest surprise: the cusps had little patches of wear. This was a reptile with tooth-to-tooth occlusion. My fossil was part mammal, part reptile.

Unbeknownst to me, Bill had unwrapped one of our blocks of rock, seen a fleck of bone, and prepared it with a needle under the microscope. None of us had known it in the field, but our expedition was a huge success. All because of Bill.

What did I learn that summer? First, I learned to listen to Chuck and Bill. Second, I learned that many of the biggest discoveries happen in the hands of fossil preparators, not in the field. As it turned out, my biggest lessons about fieldwork were yet to come.

The reptile Bill had found was a tritheledont, a creature known from South Africa as well as now from Nova Scotia. These were very rare, so we wanted to return to Nova Scotia the next summer to find more. I spent the whole winter tense with anticipation. If I could have chipped through the winter ice to find fossils, I would have done it.

In the summer of 1985, we returned to the site where we had found the tritheledont. The fossil bed was just at beach level, where a little piece of the cliff had fallen off several years before. We had to time our daily visit just so: the site was inaccessible at high tide because the water came up too high around a point we had to navigate. I'll never forget that first day of excitement when we rounded the point to find our little patch of bright orange rock. The experience was memorable for what was missing: most of the area we had worked the year before. It had weathered away the previous winter. Our lovely fossil site, containing beautiful tritheledonts, was gone with the tides.

The good news, if you could call it that, was that there was a little more orange sandstone to scan along the beach. Most of the beach, in particular the point we had to go around each morning, was made up of basalt from a 200-million-year-old lava flow. We were positive no fossils could be found there, for it is virtually axiomatic that these rocks, which were once super hot, would never preserve fossil bone. We spent five or more days timing our visits to the sites by the tides, pawing away at the orange sandstones beyond it, and finding absolutely nothing.

Our breakthrough came when the president of the local Lions Club came by our cabin one night looking for judges for the local beauty contest, to crown Parrsboro's Miss Old Home Week. The town always relied on visitors for this onerous task, because internecine passions typically run high during the event. The usual judges, an elderly couple from Quebec, were not visiting this year, and the crew and I were invited to substitute.

But in judging the beauty contest and arguing over its conclusion, we stayed up way too late, forgot about the next morning's tides, and ended up trapped around a bend in the basalt cliffs. For about two hours, we were stuck on a little promontory about fifty feet wide. The rock was volcanic and not the type one would ever choose to search for fossils. We skipped stones until we got bored, then we looked at the rocks: maybe we'd find interesting crystals or minerals. Bill disappeared around a corner, and I looked at some of the basalt behind us. After about fifteen minutes I heard my name. I'll never forget Bill's understated tone: "Uh, Neil, you might want to come over here." As I rounded the corner, I saw the excitement in Bill's eyes. Then I saw the rocks at his feet. Sticking out of the rocks were small white fragments. Fossil bones, thousands of them.

This was exactly what we were looking for, a site with small bones. It turned out that the volcanic rocks were not entirely volcanic: slivers of sandstone cut through the cliff. The rocks had been produced by an ancient mudflow associated with a volcanic eruption. The fossils were stuck in the ancient muds.

We brought tons of these rocks home. Inside were more tritheledonts, some primitive crocodiles, and other lizard-like reptiles. The tritheledonts were the gems, of course, because they showed that some kinds of reptiles already displayed our mammalian kind of chewing.

Early mammals, such as those Farish's team uncovered in Arizona, had very precise patterns of biting. Scrapes on the cusps of an upper tooth fit against mirror images of these scrapes on a lower tooth. These patterns of wear are so fine that different species of early mammals can be distinguished by their patterns of tooth wear and occlusion. Farish's Arizona mammals have a different pattern of cusps and chewing than those of the same age from South America, Europe, or China. If all we had to compare these fossils to were living reptiles, then the origin of mammalian feeding would appear to be a big mystery. As I've mentioned, crocodiles and lizards do not have any kind of matching pattern of occlusion. Here is where creatures like tritheledonts come in. When we go back in time, to rocks about 10 million years older, such as those in Nova Scotia, we find tritheledonts with an incipient version of this way of chewing. In tritheledonts, individual cusps do not interlock in a precise way, as they do in mammals; instead, the entire inner surface of the upper tooth shears against the outer surface of the lower tooth, almost like a scissors. Of course, these changes in occlusion did not happen in a vacuum. It should come as no surprise that the earliest creatures to show a mammalian kind of chewing also display mammalian features of the lower jaw, skull, and skeleton.

Because teeth preserve so well in the fossil record, we have very detailed information about how major patterns of chewing—and the ability to use new diets—arose over time. Much of the story of mammals is the story of new ways of processing food. Soon after we encounter tritheledonts in the fossil record, we start seeing all sorts of new mammal species with new kinds of teeth, as well as new ways of occluding and using them. By about 150 million years ago, in rocks from around the world, we find small rodent-size mammals with a new kind of tooth row, one that paved the way for our own existence. What made these creatures special was the complexity of their mouths: the jaw had different kinds of teeth set in it. The mouth developed a kind of division of labor. Incisors in the front became specialized to cut food, canines further back to puncture it, and molars in the extreme back to shear or mash it. These little mammals, which resemble mice, have a fundamental piece of our history inside of them. If you doubt this, imagine eating an apple lacking your incisor teeth or, better yet, a large carrot with no molars. Our diverse diet,

A tritheledont and a piece of its upper jaw discovered in Nova Scotia. Jaw fragment illustrated by Lazlo Meszoley.

ranging from fruit to meat to Twinkie, is possible only because
our distant mammalian ancestors developed a mouth with different kinds of teeth that can occlude precisely. And yes, initial
stages of this are seen in tritheledonts and other ancient relatives:
the teeth in the front have a different pattern of blades and cusps
than those in the back.

TEETH AND BONES—THE HARD STUFF

It almost goes without saying that what makes teeth special among
organs is their hardness. Teeth have to be harder than the bits of
food they break down; imagine trying to cut a steak with a sponge.
In many ways, teeth are as hard as rocks, and the reason is that
they contain a crystal molecule on the inside. That molecule,
known as hydroxyapatite, impregnates the molecular and cellular
infrastructure of both teeth and bones, making them resistant to
bending, compression, and other stresses. Teeth are extra hard
because their outer layer, enamel, is far richer in hydroxyapatite
than any other structure in the body, including bone. Enamel
gives teeth their white sheen. Of course, enamel is only one of
the layers that make up our teeth. The inner layers, such as dentine, are also filled with hydroxyapatite.

There are lots of creatures with hard tissues—clams and lobsters, for example. But they do not use hydroxyapatite; lobsters
and clams use other materials, such as calcium carbonate or chi-
tin. Also, unlike us, these animals have an exoskeleton covering the
body. Our hardness lies within.

Our particular brand of hardness, with teeth inside our mouths
and bones inside our bodies, is an essential part of who we are. We
can eat, move about, breathe, even metabolize certain minerals
because of our hydroxyapatite-containing tissues. For these capabilities, we can thank the common ancestor we share with all fish.

Every fish, amphibian, reptile, bird, and mammal on the planet is like us. All of them have hydroxyapatite-containing structures. But where did this all come from?

There is an important intellectual issue at stake here. By knowing where, when, and how hard bones and teeth came about, we will be in a position to understand why. Why did our kind of hard tissues arise? Did they come about to protect animals from their environment? Did they come about to help them move? Answers to these questions lie in the fossil record, in rocks approximately 500 million years old.

Some of the most common fossils in ancient oceans, 500 million to 250 million years old, are conodonts. Conodonts were discovered in the 1830s by the Russian biologist Christian Pander, who will reappear in a few chapters. They are small shelly organisms with a series of spikes projecting out of them. Since Pander's time, conodonts have been discovered on every continent; there are places where you cannot crack a rock without finding vast numbers of them. Hundreds of kinds of conodonts are known.

For a long time, conodonts were enigmas: scientists disagreed over whether they were animal, vegetable, or mineral. Everybody seemed to have a pet theory. Conodonts were claimed to be pieces of clams, sponges, vertebrates, even worms. The speculation ended when whole animals started to show up in the fossil record.

The first specimen that made sense of everything was found by a professor of paleontology rummaging through the basement at the University of Edinburgh: there was a slab of rock with what looked like a lamprey in it. You might recall lampreys from biology class—these are very primitive fish that have no jaws. They make their living by attaching to other fish and feeding on their bodily fluids. Embedded in the front of the lamprey impression were small fossils that looked strangely familiar. Conodonts. Other lamprey-like fossils started to come out of rocks in South Africa and later the western United States. These creatures all had an exceptional trait: they had whole assemblages of conodonts in their mouths. The conclusion became abundantly clear: conodonts were teeth. And not just any teeth. Conodonts were the teeth of an ancient jawless fish.

We had the earliest teeth in the fossil record for over 150 years before we realized what they were. The reason comes down to how fossils are preserved. The hard bits, for example teeth, tend to get preserved easily. Soft parts, such as muscle, skin, and guts, usually decay without fossilizing. We have museum cabinets full of fossil skeletons, shells, and teeth, but precious few guts and brains. On the rare occasions when we find evidence of soft tissues, they are typically preserved only as impressions or casts. Our fossil record is loaded with conodont teeth, but it took us 150 years to find the bodies. There is something else remarkable about the bodies to which conodonts belonged. They have no hard bones. These were soft-bodied animals with hard teeth.

For years, paleontologists have argued about why hard skeletons, those containing hydroxyapatite, arose in the first place. For those who believed that skeletons began with jaws, backbones, or body armor, conodonts provide an "inconvenient tooth," if you will. The first hard hydroxyapatite-containing body parts were teeth. Hard bones arose not to protect animals, but to eat them. With this, the fish-eat-fish world really began in earnest. First, big fish ate little fish; then, an arms race began. Little fish developed armor, big fish obtained bigger jaws to crack the armor, and so on. Teeth and bones really changed the competitive landscape.

Things get more interesting still as we look at some of the first animals with bony heads. As we move up in time from the earliest conodont animals, we see what the first bony-head skeletons looked like. They belonged to fish called ostracoderms, are about 500 million years old, and are found in rocks all over the world, from the Arctic to Bolivia. These fish look like hamburgers with fleshy tails.

The head region of an ostracoderm is a big disk covered by a shield of bone, looking almost like armor. If I were to open a museum drawer and show you one, you would immediately notice something odd: the head skeleton is really shiny, much like our teeth or the scales of a fish.

One of the joys of being a scientist is that the natural world has the power to amaze and surprise. Here, in ostracoderms, an obscure group of ancient jawless fish, lies a prime example. Ostra-

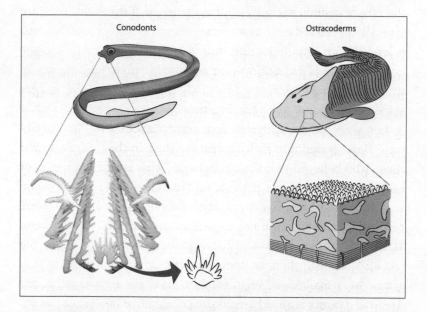

A conodont (left) and an ostracoderm (right). Conodonts were originally found iso-lated. Then, as whole animals became known, we learned that many of them func-tioned together as a tooth row in the mouths of these soft-bodied jawless fish. Ostracoderms have heads covered with a bony shield. The microscopic layers of that shield look like they are composed of little tooth-like structures. Conodont tooth row reconstruction courtesy of Dr. Mark Purnell, University of Leicester, and Dr. Philip Donoghue, University of Bristol.

coderms are among the earliest creatures with bony heads. Cut the bone of the skull open, embed it in plastic, pop it under the microscope, and you do not find just any old tissue structure; rather, you find virtually the same structure as in our teeth. There is a layer of enamel and even a layer of pulp. The whole shield is made up of thousands of small teeth fused together. This bony skull—one of the earliest in the fossil record—is made entirely of little teeth. Teeth originally arose to bite creatures; later, a version of teeth was used in a new way to protect them.

TEETH, GLANDS, AND FEATHERS

Teeth not only herald a whole new way of living, they reveal the origin of a whole new way of making organs. Teeth develop by an interaction of two layers of tissue in our developing skin. Basically, two layers approach each other, cells divide, and the layers change shape and make proteins. The outer layer spits out the molecular precursors of enamel, the inner layer the dentine and pulp of the inside of the tooth. Over time, the structure of the tooth is laid down, then tweaked to make the patterns of cusps and troughs that distinguish each species.

The key to tooth development is that an interaction between these two layers of tissue, an outer sheet of cells and an inner loose layer of cells, causes the tissue to fold and makes both layers secrete the molecules that build the organ. It turns out that exactly the same process underlies the development of all the structures that develop within skin: scales, hair, feathers, sweat glands, even mammary glands. In each case, two layers come together, fold, and secrete proteins. Indeed, the batteries of the major genetic switches that are active in this process in each kind of tissue are largely similar.

This example is akin to making a new factory or assembly

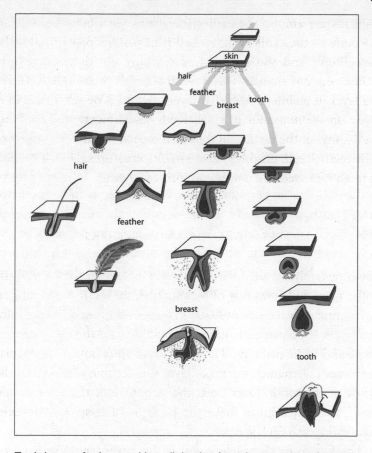

Teeth, breasts, feathers, and hair all develop from the interactions between layers of skin.

process. Once plastic injection was invented, it was used in making everything from car parts to yo-yos. Teeth are no different. Once the process that makes teeth came into being, it was modified to make the diverse kinds of organs that lie within skin. We saw this taken to a very great extreme in the ostracoderms. Birds, reptiles, and humans are just as extreme in many ways. We would never have scales, feathers, or breasts if we didn't have teeth in the first place. The developmental tools that make teeth have been repurposed to make other important skin structures. In a very real sense organs as different as teeth, feathers, and breasts are inextricably linked by history.

A theme of these first four chapters is how we can trace the same organ in different creatures. In Chapter 1 we saw that we can make predictions and find versions of our organs in ancient rocks. In Chapter 2 we saw how we can trace similar bones all the way from fish to humans. Chapter 3 shows how the real heritable part of our bodies—the DNA and genetic recipe that builds organs—can be followed in very different creatures. Here, in teeth, mammary glands, and feathers, we find a similar theme. The biological processes that make these different organs are versions of the same thing. When you see these deep similarities among different organs and bodies, you begin to recognize that the diverse inhabitants of our world are just variations on a theme.

CHAPTER FIVE

GETTING AHEAD

It was two nights before my anatomy final and I was in the lab at around two in the morning, memorizing the cranial nerves. There are twelve cranial nerves, each branching to take bizarre twists and turns through the inside of the skull. To study them, we bisected the skull from forehead to chin and sawed open some of the bones of the cheek. So there I was, holding half of the head in each hand, tracing the twisted paths that the nerves take from our brains to the different muscles and sense organs inside.

I was enraptured by two of the cranial nerves, the trigeminal and the facial. Their complicated pattern boiled down to something so simple, so outrageously easy that I saw the human head in a new way. That insight came from understanding the far simpler state of affairs in sharks. The elegance of my realization—though not its novelty; comparative anatomists had had it a century or more ago—and the pressure of the upcoming exam led me to forget where I was. At some point, I looked around. It was the middle of the night and I was alone in the lab. I also happened to be surrounded by the bodies of twenty-five human beings under sheets. For the first and last time, I got the willies. I worked myself into such a lather that the hairs on the back of my neck rose, my feet did their job, and within a nanosecond I found myself at the bus stop, out of breath. It goes without saying that I felt ridiculous. I remember telling myself: Shubin, you've become hard-core. That thought did not last long; I soon discovered I had locked my house keys in the lab.

What made me so hard-core is that head anatomy is deeply mesmerizing, in fact, beautiful. One of the joys of science is that, on occasion, we see a pattern that reveals the order in what initially seems chaotic. A jumble becomes part of a simple plan, and you feel you are seeing right through something to find its essence. This chapter is about seeing that essence inside our own heads. And, of course, the heads of fish.

THE INNER CHAOS OF THE HEAD

Head anatomy is not only complicated but hard to see, since, unlike other parts of the body, the tissues of the head are encapsulated in a bony box. We literally have to saw through the cheek, forehead, and cranium to see the vessels and organs. Having thus opened a human head, we find a clump of what looks like tangled fishing lines. Vessels and nerves make curious loops and turns as they travel through the skull. Thousands of nerve branches, muscles, and bones sit within this small box. At first glance, the whole array is a bewildering mess.

Our skulls are made up of three fundamental parts: think plates, blocks, and rods. The plates cover our brain. Pat the top of your head, and you are feeling them. These large plates fit together like jigsaw-puzzle pieces and form much of our cranium. When we were born, the plates were separate; the open spaces between them, the fontanelles, are visible in infants, occasionally throbbing with the brain tissue underneath. As we grow, the bones enlarge, and by the time we reach the age of two they have fused.

Another part of our skull lies underneath the brain, forming a platform that holds it up. Unlike the plate-like bones at the top, these bones look like complicated blocks and have many arteries and nerves running through them. The third kind of bone makes up our jaws, some bones in our ears, and other bones in our throats; these bones start development looking like rods, which ultimately break up and change shape to help us chew, swallow, and hear.

Inside the skull are a number of compartments and spaces that house different organs. Obviously, the brain occupies the largest of these. Other spaces contain our eyes, parts of our ears, and our nasal structures. Much of the challenge in understanding head anatomy comes from seeing these different spaces and organs in three dimensions.

Attached to the bones and organs in the head are the muscles we use to bite, to talk, and to move our eyes and whole head. Twelve nerves supply these muscles, each exiting the brain to travel to a different region inside our head. These are the dreaded cranial nerves.

The key to unlocking the basics of the head is to see the cranial

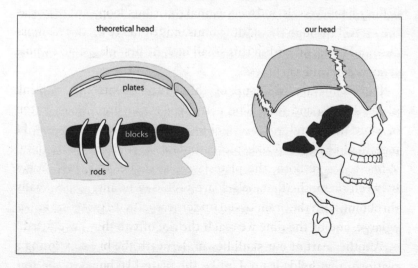

Plates, blocks, and rods: the theme for skulls. Every bone in our head can be traced to one of these things.

nerves as more than a jumble. Indeed, most of them really are simple. The simplest cranial nerves have only one function, and they attach to one muscle or organ. The cranial nerve that goes to our nasal structures, the olfactory, has one job: to take information from our nasal tissues to our brain. Some of the nerves that go to our eyes and ears are also simple in this way: the optic nerve is involved with vision; the acoustic nerve works in hearing. About four other cranial nerves only serve muscles—working to move the eyes inside the orbits, for example, or to move the head around on the neck.

But four of the cranial nerves have given medical students fits for decades. For good reason: the four have very complex functions and take tortuous paths through the head to do their jobs. The trigeminal nerve and the facial nerve deserve special mention. Both exit the brain and break up into a bewildering network of branches. Much like a cable that can carry television, Internet, and voice information, a single branch of the trigeminal or facial nerve can carry information about both sensation and action. Individual fibers for sensation and action emanate from different parts of the brain, are consolidated in cables (what we end up calling the trigeminal and facial nerves), then break up again, sending branches all over the head.

The trigeminal's branches do two major things: they control muscles, and they carry sensory information from much of our face back to our brain. The muscles controlled by the trigeminal nerve include those we use to chew as well as tiny muscles deep inside the ear. The trigeminal is also the major nerve for sensation in the face. The reason a slap to the face hurts so much, beyond the emotional pain, is because the trigeminal carries sensory information from the skin of our face back to our brain. Your dentist also knows the branches of your trigeminal nerve well. Different branches go to the roots of our teeth; a single jab of anesthetic along one of these branches can deaden the sensation of different parts of our tooth row.

The facial nerve also controls muscles and relays sensory information. As its name implies, it is the main nerve that controls the muscles of facial expression. We use these tiny muscles to smile, to frown, to raise and lower our eyebrows, to flare our nostrils, and so on. They have wonderfully evocative names. One of the major muscles that we use in frowning—it moves the corners of our mouth down—is called the depressor anguli oris. Another great name belongs to the muscle we use to furrow our brow in concern: the corrugator supercilii. Flare your nostrils and you are using your nasalis. Each of these muscles, like every other muscle of facial expression, is controlled by branches of the facial nerve. Things like an uneven smile or asymmetrically drooping eyelids are a sign that something might be wrong with the facial nerve on one side of a person's face.

You are probably beginning to see why I was staying up so late to study these nerves. Nothing about them seems to make any sense. For example, both the trigeminal and the facial nerves send tiny branches to muscles inside our ears. Why do two different nerves, which innervate entirely different parts of the face and jaw, send branches to ear muscles that lie adjacent to one another? Even more confusing, the trigeminal and facial almost crisscross as they send branches to our face and jaw. Why? With such oddly redundant functions and tortuous paths, there seems to be no rhyme or reason to their structure, much less to how these nerves match up with the plates, blocks, and rods that make up our skull.

In thinking about these nerves, I am reminded of my first days here in Chicago in 2001. I had been given space for a research laboratory in a hundred-year-old building and the lab needed new utility cables, plumbing, and air handling. I remember the day when the contractors first opened the walls to get access to the innards of the building. Their reaction to the plumbing and wiring inside my wall was almost exactly like mine when I opened the human head and saw the trigeminal and facial nerves for the first time. The wires, cables, and pipes inside the walls were a jumble. Nobody in his right mind would have designed a building from scratch this way, with cables and pipes taking bizarre loops and turns throughout the building.

And that's exactly the point. My building was constructed in 1896, and the utilities reflect an old design that has been jury-rigged further with each renovation. If you want to understand the wiring and plumbing in my building, you have to understand its history, how it was renovated for each new generation of scientists. My head has a long history also, and that history explains complicated nerves like the trigeminal and the facial.

For us, that history begins with a fertilized egg.

THE ESSENCE IN EMBRYOS

Nobody starts life with a head: sperm and egg come together to make a single cell. Between the moment of conception and the third week thereafter, we go from that single cell to a ball of cells, then to a Frisbee-shaped collection of cells, then to something that looks vaguely like a tube and includes different kinds of tissues. Between the twenty-third and twenty-eighth days after conception, the front end of the tube thickens and folds over the body, so the embryo looks as if it's already curled up in the fetal position. The head at this stage looks like a big glob. The base of this glob holds the key to much of the basic organization of our heads.

Four little swellings develop around the area that will become the throat. At about three weeks we see the first two; the other two emerge about four days later. Each swelling looks quite humble on the outside: a simple blob, separated from the next by a little crease. When you follow what happens to the blobs and creases, you begin to see the order and beauty of the head, including the trigeminal and facial nerves.

Of the cells inside each blob, known as arches, some will form bone tissue and others muscle and blood vessels. There is a complex mix of cells inside each arch; some cells divided right there while others migrated a long way to enter the arch itself. When we identify the cells in each arch according to where they end up in the adult, things start to make a lot of sense.

Ultimately, the first arch tissues form the upper and lower jaws, two tiny ear bones (the malleus and incus), and all the vessels and muscles that supply them. The second arch forms the third small ear bone (the stapes), a tiny throat bone, and most of the muscles that control facial expression. The third arch forms bones, muscles, and nerves deeper in the throat; we use these to swallow. Finally, the fourth arch forms the deepest parts of our throat, including parts of our larynx and the muscles and vessels that surround it and help it function.

If you were to shrink yourself to the size of a pinhead and travel inside the mouth of the developing embryo, you would see indentations that correspond to each swelling. There are four of these indentations. And, like the arches on the outside, cells on the indentations form important structures. The first elongates to form our Eustachian tube and some structures in the ear. The second forms the cavity that holds our tonsils. The third and fourth form important glands, including the parathyroid, thymus, and thyroid.

What I've just given you is one of the big tricks for understanding the most complicated cranial nerves and large portions of the head. When you think trigeminal nerve, think first arch. Facial nerve, second arch. The reason the trigeminal nerve goes to both the jaws and the ear is that all the structures it supplies originally developed in the first arch. The same thing is true for the facial nerve and the second arch. What do the muscles of facial expression have in common with the muscles in the ear that the facial nerve supplies? They are all second arch derivatives. As for the nerves of the third and fourth arches, their complex paths all relate to the fact that they innervate structures that arose from their respective arches. Those third and fourth arch nerves, among them the glossopharyngeal and vagus, follow the same pattern as the ones in front, each going to structures that developed from the arch they are associated with.

This fundamental blueprint of heads helps us make sense of one of the apocryphal tales in anatomy. In 1820, so the story goes, Johannes Goethe was walking through the Jewish cemetery in Venice when he spotted the decomposing skeleton of a ram. The vertebrae were exposed and above them lay a damaged skull. Goethe, in a moment of epiphany, saw that the breaks in the skull made it look like a gnarled mess of vertebrae. To Goethe, this revealed the essential pattern within: the head is made up of ver-

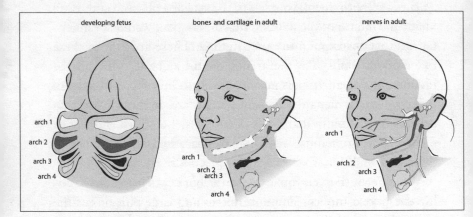

If we follow the gill arches from an embryo to an adult, we can trace the origins of jaws, ears, larynx, and throat. Bones, muscles, nerves, and arteries all develop inside these gill arches.

tebrae that fused and grew a vault to hold our brains and sense organs. This was a revolutionary idea because it linked heads and bodies as two versions of the same fundamental plan. The notion must have been in the drinking water in the early 1800s because other people, among them Lorenz Oken, allegedly came up with virtually the same idea in a similar setting.

Goethe and Oken were both picking up something very profound, although they could not have known it at the time. Our body is segmented, and this pattern is most clearly seen in our vertebrae. Each vertebra is a block that represents a segment of our body. The organization of our nerves is also segmental, correlating closely with the pattern of the vertebrae. Nerves exit the spinal cord to supply the body. The segmental configuration is obvious when you look at the levels of the spinal cord that are associated with each part of our body. For example, the muscles in our legs are supplied by nerves that exit from lower parts of the spinal cord than those that supply our arms. Heads may not look it, but they also contain a very deep segmental pattern. Our arches define segments of bones, muscles, arteries, and nerves. Look in the adult, and you won't see this pattern. We see it only in the embryo.

Our skulls lose all overt evidence of their segmental origins as we go from embryo to adult. The plate-like bones of our skulls form over our gill arches, and the muscles, nerves, and arteries, which all had a very simple segmental pattern early on, are rewired to make our adult heads.

Knowing something about development can help us predict where to look for what is missing in children who have certain birth defects. For example, children born with first arch syndrome have a tiny jaw and nonfunctioning ears with no malleus or incus bone. Missing are structures that normally would have formed from the first arch.

The arches are the road map for major chunks of the skull,

from the most complicated cranial nerves to the muscles, arteries, bones, and glands inside. The arches are also a guide to something else: our very deep connection with sharks.

OUR INNER SHARK

The take-home message of many a lawyer joke is that lawyers are an especially voracious kind of shark. Teaching embryology during one of the recurring vogues for these jokes, I remember thinking that the joke is on all of us. We're all modified sharks—or, worse, there is a lawyer inside each of us.

As we've seen, much of the secret of heads lies in the arches, the swellings that gave us the road map for the complicated cranial nerves and key structures inside the head. Those insignificant-looking swellings and indentations have captured the imagination of anatomists for 150 years, because they look like the gill slits in the throat regions of fish and sharks.

Fish embryos have these bulges and indentations, too. In fish, the indentations ultimately open up to form the spaces between the gills where water flows. In us, the indentations normally seal over. In abnormal cases, gill slits fail to close and remain open as pouches or cysts. A branchial cyst, for example, is often a benign fluid-filled cyst that forms in an open pouch inside the neck; the pouch is created by the failure of the third or fourth arch to close. Rarely, children are born with an actual vestige of an ancient gill arch cartilage, a little rod that represents a gill bar from the third arch. In these instances, my surgical colleagues are operating on an inner fish that unfortunately has come back to bite us.

Every head on every animal from a shark to a human shares those four arches in development. The richness of the story lies in what happens inside each arch. Here, we can make a point-by-point comparison between our heads and those of sharks.

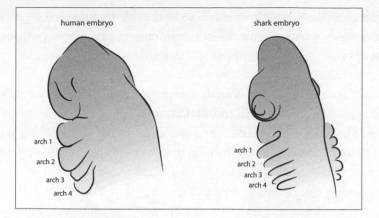

The gill region of a developing human and a developing shark look the same early on.

Look at the first arch in a human and a shark, and you find a very similar state of affairs: jaws. The major difference is that humans' first arch also forms some ear bones, which we do not see in sharks. Unsurprisingly, the cranial nerve that supplies the jaws of humans and sharks is the first arch nerve, the trigeminal nerve.

The cells inside the second gill arch divide, change, and give rise to a bar of cartilage and muscle. In us, the bar of cartilage breaks up to form one of the three bones of our middle ear (the stapes) and some other small structures at the base of the head and throat. One of these bones, called the hyoid, assists us in swallowing. Take a gulp, listen to music, and thank the structures that form from your second arch.

In a shark, the second arch rod breaks up to form two bones that support the jaws: a lower one that compares with our hyoid and an upper one that supports the upper jaw. If you have ever watched a great white shark try to chomp something—a diver in a cage, for example—you have probably noticed that the upper jaw can extend and retract as the shark bites. The upper bone of this second arch is part of the lever system of bones that rotate to make that possible. That upper bone is remarkable in another way, too. It compares with one of the bones in our middle ear: the stapes. Bones that support the upper and lower jaws in sharks are used in us to swallow and hear.

As for the third and fourth arches, we find that many of the structures we use to talk and swallow are, in sharks, parts of tissues that support the gills. The muscles and cranial nerves we use to swallow and talk move the gills in sharks and fish.

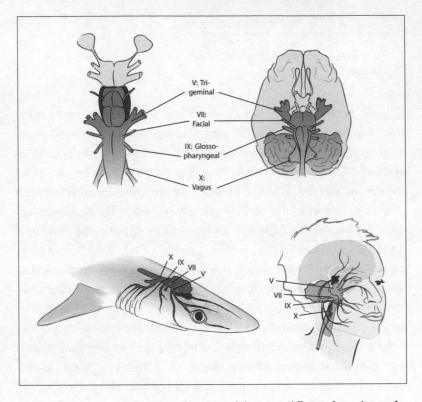

At first glance, our cranial nerves (bottom right) appear different from those of a shark (bottom left). But look closely and you will find profound similarities. Virtually all of our nerves are present in sharks. The parallels go deeper still: equivalent nerves in sharks and humans supply similar structures, and they even exit the brain in the same order (top left and right).

Our head may look incredibly complicated, but it is built from a simple and elegant blueprint. There is a pattern common to every skull on earth, whether it belongs to a shark, a bony fish, a salamander, or a human. The discovery of this pattern was a major accomplishment of nineteenth-century anatomy, a time when anatomists were putting embryos of all kinds of species under the microscope. In 1872, the Oxford anatomist Francis Maitland Balfour first saw the basic plan of heads when he looked at sharks and saw the bulges, the gill arches, and the structures inside. Unfortunately, he died soon after in a mountaineering accident in the Swiss Alps. He was only in his thirties.

GILL ARCH GENES

During the first three weeks after conception, whole batteries of genes are turned on and off in our gill arches and throughout the tissues that will become our future brain. These genes instruct cells to make the different portions of our head. Think of each region of our head as gaining a genetic address that makes it distinctive. Modify this genetic address and we can modify the kinds of structures that develop there.

For example, a gene known as *Otx* is active in the front region, where the first gill arch forms. Behind it, toward the back of the head, a number of so-called *Hox* genes are active. Each gill arch has a different complement of *Hox* genes active in it. With this information, we can make a map of our gill arches and the constellation of genes active in making each.

Now we can do experiments: change the genetic address of one gill arch into that of another. Take a frog embryo, turn off some genes, make the genetic signals similar in the first and second arches, and you end up with a frog that has two jaws: a mandible develops where a hyoid bone would normally be. This shows the power of the genetic addresses in making our gill arches. Change the address, and you change the structures in the arch. The power of this approach is that we can now experiment with the basic design of heads: we can manipulate the identity of the gill arches almost at will, by changing the activity of the genes inside.

TRACING HEADS: FROM HEADLESS WONDERS TO OUR HEADED ANCESTORS

Why stop at frogs and sharks? Why not extend our comparison to other creatures, like insects or worms? But why would we do this when none of these creatures has a skull, much less cranial nerves? None of them even has bones. When we leave fish for worms, we get to a very soft and headless world. Bits of ourselves are there, though, if you look closely.

Those of us who teach comparative anatomy to undergraduates usually begin the course with a slide of *Amphioxus*. Every September, hundreds of *Amphioxus* slides appear on screens in college lecture halls from Maine to California. Why? Remember the simple dichotomy between invertebrates and vertebrates? *Amphioxus* is a worm, an invertebrate, that shares many features with backboned animals such as fish, amphibians, and mammals. *Amphioxus* lacks a backbone, but like all creatures with backbones, it has a nerve cord that runs along its back. In addition, a rod runs the length of its body, parallel to the nerve cord. This rod, known as the notochord, is filled with a jelly-like substance and provides support for the body. As embryos, we have a notochord, too, but unlike *Amphioxus*'s, ours breaks up and ultimately becomes part of the disks that lie between our vertebrae. Rupture a disk and the jelly-like substance of what was once a notochord can wreak havoc when it pinches nerves or interferes with the ability of one verte-

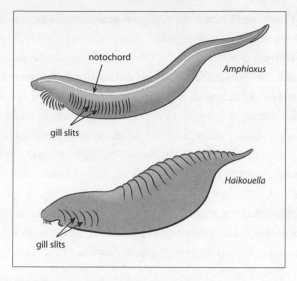

The closest relatives to animals with heads are worms with gill slits. Shown are *Amphioxus* and a reconstruction of a fossil worm (*Haikouella*) over 530 million years old. Both worms have a notochord, a nerve cord, and gill slits. The fossil worm is known from over three hundred individual specimens from southern China.

bra to move along the next. When we injure a disk, a very ancient part of our body plan is rupturing. Thanks a lot, *Amphioxus*.

Amphioxus is not unique among worms. Some of the best examples are not in the oceans of today but in ancient rocks of China and Canada. Buried in sediments over 500 million years old are small worms that lack heads, complex brains, or cranial nerves. They may not look like much, being small smudges in the rock, but the preservation of these fossils is incredible. When you look under a microscope, you find beautifully preserved impressions that display their soft anatomy in fine detail, occasionally even with impressions of skin. They show something else wonderful, too. They are the earliest creatures with notochords and nerve

cords. These worms are telling us something about the origin of
parts of our bodies.

But there is something else we share with these little worms:
gill arches. *Amphioxus,* for example, has them in abundance, and
associated with each arch is a little bar of cartilage. Like the cartilages that form our jaws, our ear bones, and parts of our voice box,
these rods support the gill slit. The essence of our head goes back
to worms, organisms that do not even have a head. What does
Amphioxus do with the gill arches? It pumps water through them
to filter out little particles of food. From so humble a beginning
comes the basic structures of our own head. Just as teeth, genes,
and limbs have been modified and their functions repurposed
over the ages, so, too, has the basic structure of our head.

CHAPTER SIX

THE BEST-LAID (BODY) PLANS

We are a package of about two trillion cells assembled in a very precise way. Our bodies exist in three dimensions, with our cells and organs in their proper places. The head is on top. The spinal cord is toward our back. Our guts are on the belly side. Our arms and legs are to the sides. This basic architecture distinguishes us from primitive creatures organized as clumps or disks of cells.

The same design is also an important part of the bodies of other creatures. Like us, fish, lizards, and cows have bodies that are symmetrical with a front/back, top/bottom, and left/right. Their front ends (corresponding to the top of an upright human) all have heads, with sense organs and brains inside. They have a spinal cord that runs the length of the body along the back. Also like us, they have an anus, which is at the opposite end of their bodies from the mouth. The head is on the forward end, in the direction they typically swim or walk. As you can imagine, "anus-forward" wouldn't work very well in most settings, particularly aquatic ones. Social situations would be a problem, too.

It is more difficult to find our basic design in really primitive animals—jellyfish, for example. Jellyfish have a different kind of body plan: their cells are organized into disks that have a top and bottom. Lacking a front and back, a head and tail, and a left and right, jellyfish body organization appears very different from our own. Don't even bother trying to compare your body plan with a sponge. You could try, but the mere fact that you were trying would reveal something more psychiatric than anatomical.

To properly compare ourselves with these primitive animals, we need some tools. Just as with heads and limbs, our history is written within our development from egg to adult. Embryos hold the clues to some of the profound mysteries of life. They also have the ability to derail my plans.

THE COMMON PLAN: COMPARING EMBRYOS

I entered graduate school to study fossil mammals and ended up three years later studying fish and amphibians for my dissertation. My fall from grace, if you want to call it that, happened when I started to look at embryos. We had a lot of embryos in the lab: salamander larvae, fish embryos, even fertilized chicken eggs. I'd routinely pop them under the microscope to see what was going on. The embryos of all the species looked like little whitish batches of cells, no more than an eighth of an inch long. It was exciting watching development progress; as the embryo got bigger, the yolk, its food supply, got smaller and smaller. By the time the yolk was gone, the embryo was usually big enough to hatch.

Watching the process of development brought about a huge intellectual transformation in me. From such simple embryonic beginnings—small blobs of cells—came wonderfully complex birds, frogs, and trout comprising trillions of cells arranged in just the right way. But there was more. The fish, amphibian, and chicken embryos were like nothing I had ever seen before in biology. They all looked generally alike. All of them had a head with gill arches. All of them had a little brain that began its development with three swellings. All of them had little limb buds. In fact, the limbs were to become my thesis, the focus of my next three years' work. Here, in comparing how the skeleton developed

in birds, salamanders, frogs, and turtles,
as different as bird wings and frog legs l
their development. In seeing these embi
mon architecture. The species ended u
they started from a generally similar pl:
it almost seems that the differences :
amphibians, and fish simply pale in com
mental similarities. Then I learned of the work of Karl Ernst von
Baer.

In the 1800s, some natural philosophers looked to embryos to
try to find the common plan for life on earth. Paramount among
these observers was Karl Ernst von Baer. Born to a noble family,
he initially trained to be a physician. His academic mentor suggested that he study chicken development and try to understand
how chicken organs developed.

Unfortunately, von Baer could not afford incubators to work on
chickens, nor could he afford many eggs. This was not very promising. Lucky for him, he had an affluent friend, Christian Pander,
who could afford to do the experiments. As they looked at embryos,
they found something fundamental: *all organs in the chicken can be
traced to one of three layers of tissue in the developing embryo.* These
three layers became known as the germ layers. They achieved
almost legendary status, which they retain even to this day.

Pander's three layers gave von Baer the means to ask important
questions. Do all animals share this pattern? Are the hearts, lungs,
and muscles of all animals derived from these layers? And, importantly, do the same layers develop into the same organs in different
species?

Von Baer compared the three layers of Pander's chicken embryos with everything else he could get his hands on: fish, reptiles,
and mammals. Yes, every animal organ originated in one of these
three layers. Significantly, the three layers formed the same structures in every species. Every heart of every species formed from

r. Another layer gave rise to every brain of every ani-
so on. No matter how different the species look as
as tiny embryos they all go through the same stages of
elopment.

To fully appreciate the importance of this, we need to look again at our first three weeks after conception. At the moment of fertilization, major changes happen inside the egg—the genetic material of the sperm and egg fuses and the egg begins to divide. Ultimately, the cells form a ball. In humans, over about five days, the single-cell body divides four times, to produce a ball of sixteen cells. This ball of cells, known as a blastocyst, resembles a fluid-filled balloon. A thin spherical wall of cells surrounds some fluid in the center. At this "blastocyst stage" there still does not appear to be any body plan—there is no front and back, and certainly there are not yet any different organs or tissues. On about the sixth day after conception, the ball of cells attaches to its mother's uterus and begins the process of connecting to it so that mother and embryo can join bloodstreams. There is still no evidence of the body plan. It is a far cry from this ball of cells to anything that you'd recognize as any mammal, reptile, or fish, much less a human.

If we are lucky, our ball of cells has implanted in our mother's uterus. When a blastocyst implants in the wrong place—when there is an "ectopic implantation"—the results can be dangerous. About 96 percent of ectopic implantations happen in the uterine (or fallopian) tubes, near where conception happens. Sometimes mucus blocks the easy passage of the blastocyst to the uterus, causing it to implant improperly in the tubes. Ectopic pregnancy can cause various tissue ruptures if not caught in time. In really rare cases, the blastocyst is expelled into the mother's body cavity, the space between her guts and body wall. In even rarer cases, these blastocysts will implant on the outside lining of the mother's rectum or uterus and the fetus develops to full term! Although these fetuses can sometimes be delivered by an abdominal incision, such implantation is generally very dangerous because it increases the risk of maternal death by bleeding by a factor of 90, as compared with a normal implantation inside the uterus.

In any event, at this stage of development we are extremely humble-looking creatures. Around the beginning of our second week after conception, the blastocyst has implanted, with one part of the ball embedded in the wall of the uterus, and the other free. Think of a balloon pushed into a wall: this flattened disk becomes the human embryo. Our *entire* body forms from only the top part of this ball, the part that is mushed into the wall. The part of the blastocyst below the disk covers the yolk. At this stage of development, we look like a Frisbee, a simple two-layered disk.

How does this oval Frisbee end up with von Baer's three germ layers and go on to look anything like a human? First, cells divide and move, causing tissues to fold in on themselves. Eventually, as tissues move and fold, we become a tube with a folded swelling at the head end and another at the tail. If we were to cut ourselves in half right about now, we would find a tube within a tube. The outer tube would be our body wall, the inner tube our eventual digestive tract. A space, the future body cavity, separates the two tubes. This tube-within-a-tube structure stays with us our entire lives. The gut tube gets more complicated, with a big sac for a stomach and long intestinal twists and turns. The outer tube is complicated by hair, skin, ribs, and limbs that push out. But the basic plan persists. We may be more complicated than we were at twenty-one days after conception, but we are still a tube within a tube, and all of our organs derive from one of the three layers of tissue that appeared in our second week after conception.

The names of these three all-important layers are derived from their position: the outer layer is called ectoderm, the inner layer endoderm, and the middle layer mesoderm. Ectoderm forms much of the outer part of the body (the skin) and the nervous system.

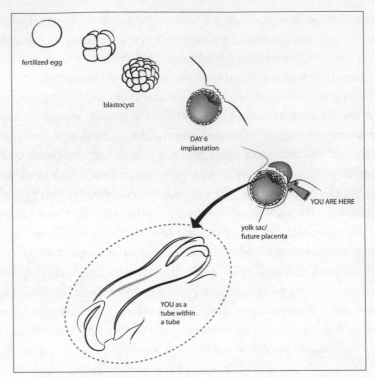

fertilized egg

blastocyst

DAY 6
implantation

YOU ARE HERE

yolk sac/
future placenta

YOU as a
tube within
a tube

Our early days, the first three weeks after conception. We go from being a single cell to a ball of cells and end up as a tube.

Endoderm, the inside layer, forms many of the inner structures of the body, including our digestive tract and numerous glands associated with it. The middle layer, the mesoderm, forms tissue in between the guts and skin, including much of our skeleton and our muscles. Whether the body belongs to a salmon, a chicken, a frog, or a mouse, all of its organs are formed by endoderm, ectoderm, and mesoderm.

Von Baer saw how embryos reveal fundamental patterns of life. He contrasted two kinds of features in development: features shared by every species, and features that vary from species to species. Features such as the tube-within-a-tube arrangement are shared by all animals with a backbone: fish, amphibians, reptiles, birds, and mammals. These common features appear relatively early in development. The features that distinguish us—bigger brains in humans, shells on turtles, feathers on birds—arise relatively later.

Von Baer's approach is very different from the "ontogeny recapitulates phylogeny" idea you might have learned in school. Von Baer simply compared embryos and noted that the embryos of different species looked more similar to each other than do the adults of those species. The "ontogeny recapitulates phylogeny" approach championed decades later by Ernst Haeckel made the claim that each species tracked its evolutionary history as it proceeded through development. Accordingly, the embryo of a human went through a fish, a reptile, and a mammal stage. Haeckel would compare a human embryo to an adult fish or a lizard. The

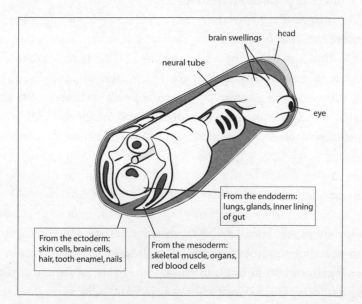

At four weeks after conception, we are a tube within a tube and have the three germ layers that give rise to all our organs.

differences between the ideas of von Baer and Haeckel might seem subtle, but they are not. In the past one hundred years, time and new evidence have treated von Baer much more kindly. In comparing embryos of one species to adults of another, Haeckel was comparing apples to oranges. A more meaningful comparison is one where we can ultimately uncover the mechanisms that drive evolution. For that, we compare embryos of one species to embryos of another. The embryos of different species are not completely identical, but their similarities are profound. All have gill arches, notochords, and look like a tube within a tube at some stage of their development. And, importantly, embryos as distinct as fish and people have Pander and von Baer's three germ layers.

All of these comparisons lead us to the real issue at stake. How does the embryo "know" to develop a head at the front end and an anus at the back? What mechanisms drive development and make cells and tissues able to form bodies?

To answer these questions required a whole new approach. Rather than simply comparing embryos as in von Baer's day, we had to find a new way of analyzing them. The latter part of the nineteenth century ushered in the era, which we first discussed in Chapter 3, when embryos were chopped, grafted, split, and treated with virtually every kind of chemical imaginable. All in the name of science.

EXPERIMENTING WITH EMBRYOS

Biologists at the turn of the twentieth century were grappling with fundamental questions about bodies. Where in the embryo does the information to build them lie? Is this information contained in every cell or in patches of cells? And what form does this information take—is it a special kind of chemical?

Beginning in 1903, the German embryologist Hans Spemann

began to investigate how cells learned to build bodies during development. His goal was to find where the body-building information resides. The big question for Spemann was whether all the cells in the embryo have enough information to build whole bodies, or whether that information is confined to certain parts of the developing embryo.

Working with newt eggs, which were easy to obtain and relatively easy to fiddle with in the lab, Spemann devised a clever experiment. He cut off a strand of his infant daughter's hair and made a miniature lasso out of it. Baby hair is remarkable stuff; soft, thin, and pliant, it made the ideal material for tying up a tiny sphere such as a newt egg. Spemann did exactly that to a developing newt egg, pinching one side off from the other. Manipulating the nuclei of the cells a bit, he let the resulting contraption develop and watched what happened. The embryo formed twins: two complete salamanders emerged, each with a normal body plan and each entirely viable. The conclusion was obvious: from one egg can come more than one individual. This is what identical twins are. Biologically, Spemann had demonstrated that in the early embryo some cells have the capacity to form a whole new individual on their own.

This experiment was only the beginning of a whole new phase of discovery.

In the 1920s Hilde Mangold, a graduate student in Spemann's laboratory, started to work with small embryos. The fine control she had of her fingers made her able to do some incredibly demanding experiments. At the stage of development with which Mangold worked, the salamander embryo is a sphere about a sixteenth of an inch in diameter. She lopped off a tiny piece of tissue, smaller than a pinhead, from one part of the embryo and grafted it onto the embryo of another species. What Mangold transplanted wasn't just any patch, but an area where cells that were to form much of the three germ layers were moving and folding. Mangold

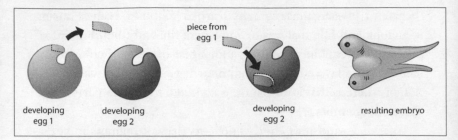

piece from
egg 1

developing   developing   developing   resulting embryo
egg 1   egg 2   egg 2

Just by moving a small patch of tissue in the embryo, Mangold produced twins.

was so skilled that the grafted embryos actually continued to develop, giving her a pleasant surprise. The grafted patch led to the formation of a whole new body, including a spinal cord, back, belly, even a head.

Why is all this important? Mangold had discovered a small patch of tissue that was able to direct other cells to form an entire body plan. The tiny, incredibly important patch of tissue containing all this information was to be known as the Organizer.

Mangold's dissertation work was ultimately to win the Nobel Prize, but not for her. Hilde Mangold died tragically (the gasoline stove in her kitchen caught fire) before her thesis could even be published. Spemann won the Nobel Prize in Medicine in 1935, and the award cites "his discovery of the Organizer and its effect in embryonic development."

Today, many scientists consider Mangold's work to be the single most important experiment in the history of embryology.

At roughly the same time that Mangold was doing experiments in Spemann's lab, W. Vogt (also in Germany) was designing clever techniques to label cells, or batches of them, and thus allow the experimenter to watch what happens as the egg develops. Vogt was able to produce a map of the embryo that shows where every organ originates in the egg. We see the antecedents of the body plan in the cell fates of the early embryo.

From the early embryologists, people like von Baer, Pander, Mangold, and Spemann, we have learned that all the parts of our adult bodies can be mapped to individual batches of cells in the simple three-layered Frisbee, and the general structure of the body is initiated by the Organizer region discovered by Mangold and Spemann.

Cut, slice, and dice, and you'll find that all mammals, birds, amphibians, and fish have Organizers. You can even sometimes swap one species' Organizer for another. Take the Organizer region from a chicken and graft it to a salamander embryo: you get a twinned salamander.

But just what is an Organizer? What inside it tells cells how to build bodies? DNA, of course. And it is in this DNA that we will find the inner recipe that we share with the rest of animal life.

OF FLIES AND MEN

Von Baer watched embryos develop, compared one species to another, and saw fundamental patterns in bodies. Mangold and Spemann physically distorted embryos to learn how their tissues build bodies. In the DNA age, we can ask questions about our own genetic makeup. How do our genes control the development of our tissues and our bodies? If you ever thought that flies are unimportant, consider this: mutations in flies gave us important clues to the major body plan genes active in *human* embryos. We put this kind of thinking to use in the discovery of genes that build fingers and toes. Now we'll see how it tells us about the ways entire bodies are built.

Flies have a body plan. They have a front and a back, a top and a bottom, and so on. Their antennae, wings, and other appendages pop out of the body in the right place. Except when they don't. Some mutant flies have limbs growing out of their heads. Others have duplicate wings and extra body segments. These are among the fly mutants that tell us why our vertebrae change shape from the head end to the anal end of the body.

People have been studying abnormal flies for over a hundred years. Mutants with one particular kind of abnormality got special attention. These flies had organs in the wrong places—a leg where an antenna should have been; an extra set of wings—or were missing body segments. Something was messing with their fundamental body plan. Ultimately, these mutants arise from some sort of error in the DNA. Remember that genes are stretches of DNA that lie on the chromosome. Using a variety of techniques that allow us to visualize the chromosome, we can identify the patch of the chromosome responsible for the mutant effect. Essentially, we breed mutants to make a whole population where every individual has the genetic error. Then, using a variety of molecular markers, we compare the genes of individuals with the mutation to those without. This allows us to pinpoint the region and the likely stretch of chromosome responsible for the mutant effect. It turns out that a fly has eight genes that make such mutants. These genes lie next to one another on one of the long DNA strands of the fly. The genes that affect the head segments lie next to those that affect the segments in the middle of the fly, the part of the body that contains the wings. These bits of DNA, in turn, lie adjacent to the ones that control the development of the rear part of the fly. There is a wonderful order to the way the genes are organized: their position along the DNA strand parallels the structure of the body from front to back.

Now the challenge was to identify the structure of the DNA actually responsible for the mutation. Mike Levine and Bill McGinnis, in Walter Gehring's lab in Switzerland, and Matt Scott, in Tom Kauffman's lab in Indiana, noticed that in the middle of each gene was a short DNA sequence that was virtually identical in each species they looked at. This little sequence is called a homeobox. The eight genes that contain the homeobox are called *Hox* genes.

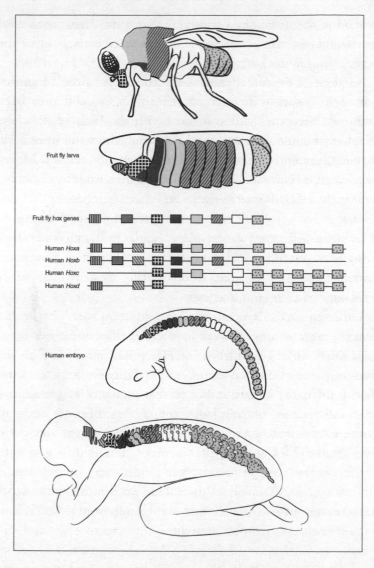

Fruit fly larva

Fruit fly hox genes

Human *Hoxa*
Human *Hoxb*
Human *Hoxc*
Human *Hoxd*

Human embryo

Hox genes in flies and people. The head-to-tail organization of the body is under the control of different *Hox* genes. Flies have one set of eight hox genes, each represented as a little box in the diagram. Humans have four sets of these genes. In flies and people, the activity of a gene matches its position on the DNA: genes active in the head lie at one end, those in the tail at another, with genes affecting the middle of the body lying in between.

When the scientists fished around for this gene sequence in other species, they found something so uniform that it came as a true surprise: *versions of the* Hox *genes appear in every animal with a body.*

Versions of the same genes sculpt the front-to-back organization of the bodies of creatures as different as flies and mice. Mess with the *Hox* genes and you mess with the body plan in predictable ways. If you make a fly that lacks a gene active in a middle segment, the midsection of the fly is missing or altered. Make a mouse that lacks one of the genes that specifies thoracic segments, and you transform parts of the back.

Hox genes also establish the proportions of our bodies—the sizes of the different regions of our head, chest, and lower back. They are involved in the development of individual organs, limbs, genitalia, and guts. Changes in them bring about changes in the ways our bodies are put together.

Different kinds of creatures have different numbers of *Hox* genes. Flies and other insects have eight, mice and other mammals thirty-nine. The thirty-nine *Hox* genes in mice are all versions of the ones that are found in flies. This similarity has led to the idea that the large number of mammalian *Hox* genes arose from a duplication of the smaller complement of genes in the fly. Despite these differences in number, the mouse genes are active from front to back in a very precise order just as the fly genes are.

Can we go even deeper in our family tree, finding similar stretches of DNA involved in making even more fundamental parts of our bodies? The answer, surprisingly, is yes. And it links us to animals even simpler than flies.

DNA AND THE ORGANIZER

At the time when Spemann won the Nobel Prize, the Organizer was all the rage. Scientists sought the mysterious chemical that could induce the entire body plan. But just as popular culture has yo-yos and Tickle Me Elmo dolls, so science has fads that wax and wane. By the 1970s, the Organizer was viewed as little more than a curiosity, a clever anecdote in the history of embryology. The reason for this fall from grace was that no one could decipher the mechanisms that made it work.

The discovery of *Hox* genes in the 1980s changed everything. In the early 1990s, when the Organizer concept was still decidedly unfashionable, Eddie De Robertis's laboratory at UCLA was looking for *Hox* genes in frogs, using techniques like Levine and McGinnis's. The search was broad and it netted many different kinds of genes. One of these had a very special pattern of activity. It was active at the exact site in the embryo that contains the Organizer, and it was active at exactly the right time of development. I can only imagine what De Robertis felt when he found that gene. He was looking at the Organizer, and there in the Organizer was a gene that seemed specifically to control it or be linked to its activity in the embryo. The Organizer was back.

Organizer genes started popping up in laboratories everywhere. While doing a different kind of experiment, Richard Harland at Berkeley found another gene, which he called *Noggin*. *Noggin* does exactly what an Organizer gene should. When Harland took some *Noggin* and injected it into the right place in an embryo, it functioned exactly like the Organizer. The embryo developed two body axes, including two heads.

Are De Robertis's gene and *Noggin* the actual bits of DNA that make up the Organizer? The answer is yes and no. Many genes, including these two, interact to organize the body plan. Such systems are complex, because genes can play many different roles during development. *Noggin*, for example, plays a role in the development of the body axis but is also involved with a host of other organs. Furthermore, genes do not act alone to specify complicated cell behaviors like those we see in head development.

Genes interact with other genes at all stages of development. One gene may inhibit the activity of another or promote it. Sometimes many genes interact to turn another gene on or off. Fortunately, new tools allow us to study the activity of thousands of genes in a cell at once. Couple this technology with new computer-based ways of interpreting gene function and we have enormous potential to understand how genes build cells, tissues, and bodies.

Understanding these complex interactions between batteries of genes sheds light on the actual mechanisms that build bodies. *Noggin* serves as a great example. *Noggin* alone does not instruct any cell in the embryo about its position on the top–bottom axis; rather, it acts in concert with several other genes to do this. Another gene, *BMP-4*, is a bottom gene; it is turned on in cells that will make the bottom, or belly side, of an embryo. There is an important interaction between *BMP-4* and *Noggin*. Wherever *Noggin* is active, *BMP-4* cannot do its job. The upshot is that *Noggin* does not tell cells to develop as "cells on the top of the body"; instead, it turns off the signal that would make them *bottom* cells. These off-on interactions underlie virtually all developmental processes.

AN INNER SEA ANEMONE

It is one thing to compare our bodies with those of frogs and fish. In a real sense we and they are much alike: we all have a backbone, two legs, two arms, a head, and so on. What if we compare ourselves with something utterly different, for example jellyfish and their relatives?

Most animals have body axes defined by their direction of movement or by where their mouth and anus lie relative to each other. Think about it: our mouth is on the opposite end of the body from our anus and, as in fish and insects, it is usually in the direction "forward."

How can we try to see ourselves in animals that have no nerve cord at all? How about no anus and no mouth? Creatures like jellyfish, corals, and sea anemones have a mouth, but no anus. The opening that serves as a mouth also serves to expel waste. While that odd arrangement may be convenient for jellyfish and their relatives, it gives biologists vertigo when they try to compare these creatures to anything else.

A number of colleagues, Mark Martindale and John Finnerty among them, have dived into this problem by studying the development of this group of animals. Sea anemones have been remarkably informative, because they are close relatives of jellyfish and they have a very primitive body pattern. Also, sea anemones have a very unusual shape, one that at first glance would seem to make them worthless as a form to compare to us. A sea anemone is shaped like a tree trunk with a long central stump and a bunch of tentacles at the end. This odd shape makes it particularly appealing, since it might have a front and a back, a top and a bottom. Draw a line from the mouth to the base of the animal. Biologists have given that line a name: the oral–aboral axis. But naming it doesn't make it more than an arbitrary line. If it *is* real, then its development should resemble the development of one of our own body dimensions.

Martindale and his colleagues discovered that primitive versions of some of our major body plan genes—those that determine our head-to-anus axis—are indeed present in the sea anemone. And, more important, these genes are active along the oral–aboral axis. This in turn means that the oral–aboral axis of these primitive creatures is genetically equivalent to our head-to-anus axis.

One axis down, another to go. Do sea anemones have anything

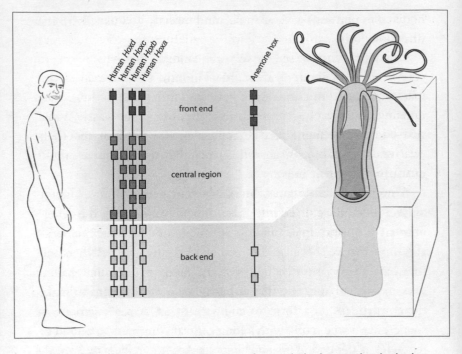

Jellyfish relatives, such as sea anemones, have a front and a back as we do, a body plan set up by versions of the same genes.

analogous to our belly-to-back axis? Sea anemones don't seem to have anything comparable. Despite this, Martindale and his colleagues took the bold step of searching in the sea anemone for the genes that specify our belly-to-back axis. They knew what our genes looked like, and this gave them a search image. They uncovered not one, but many different belly-to-back genes in the sea anemone. But although these genes were active along an axis in the sea anemone, that axis didn't seem to correlate with any pattern in how the adult animal's organs are put together.

Just what this hidden axis could be is not apparent from the outside of the animal. If we cut one in half, however, we find an important clue, another axis of symmetry. Called the directive axis, it seems to define two distinct sides of the creature, almost a left and a right. This obscure axis was known to anatomists back in the 1920s but remained a curiosity in the scientific literature. Martindale, Finnerty, and their team changed that.

All animals are the same but different. Like a cake recipe passed down from generation to generation—with enhancements to the cake in each—the recipe that builds our bodies has been passed down, and modified, for eons. We may not look much like sea anemones and jellyfish, but the recipe that builds us is a more intricate version of the one that builds them.

Powerful evidence for a common genetic recipe for animal bodies is found when we swap genes between species. What happens when you swap a body-building gene from an animal that has a complex body plan like ours with one from a sea anemone? Recall the gene *Noggin,* which in frogs, mice, and humans is turned on in places that will develop into back structures. Inject extra amounts of frog Noggin into a frog egg, and the frog will grow extra back structures, sometimes even a second head. In sea anemone embryos, a version of *Noggin* is also turned on at one end of the directive axis. Now, the million-dollar experiment: take the product of Noggin from a sea anemone and inject it into a frog embryo. The result: a frog with extra back structures, almost the same result as if the frog were injected with its own Noggin.

Now, though, as we go back in time, we are left with what looks like a huge gap. Everything in this chapter had a body. How do we compare ourselves with things that have no bodies at all—with single-celled microbes?

CHAPTER SEVEN

ADVENTURES IN BODYBUILDING

When I wasn't out in the field collecting fossils, much of my graduate career was spent staring into a microscope, looking at how cells come together to make bones.

I would take the developing limb of a salamander or a frog, and stain the cells with dyes that turn developing cartilage blue and bones red. I could then make the rest of the tissues clear by treating the limb with glycerin. These were beautiful preparations: the embryo entirely clear and all the bones radiating the colors of the dyes. It was like looking at creatures made of glass.

During these long hours at the microscope, I was literally watching an animal being built. The earliest embryos would have tiny little limb buds and the cells inside would be evenly spaced. Then, at later stages, the cells would clump inside the limb bud. In successively older embryos, the cells would take different shapes and the bones would form. Each of those clumps I saw during the early stages became a bone.

It is hard not to feel awestruck watching an animal assemble itself. Just like a brick house, a limb is built by smaller pieces joining to make a larger structure. But there is a huge difference. Houses have a builder, somebody who actually knows where all the bricks need to go; limbs and bodies do not. The information that builds limbs is not in some architectural plan but is contained within each cell. Imagine a house coming together spontaneously from all the information contained in the bricks: that is how animal bodies are made.

Much of what makes a body is locked inside the cell; in fact, much of what makes us unique is there, too. Our body looks different from that of a jellyfish because of the ways our cells attach to one another, the ways they communicate, and the different materials they make.

Before we could even have a "body plan"—let alone a head, brain, or arm—there had to be a way to make a body in the first place. What does this mean? To make all of a body's tissues and structures, cells had to know how to cooperate—to come together to make an entirely new kind of individual.

To understand the meaning of this, let's first consider what a body is. Then, let's address the three great questions about bodies: When? How? And Why? When did bodies arise, how did they come about, and, most important, why are there bodies at all?

HABEAS CORPUS: SHOW ME THE BODY

Not every clump of cells can be awarded the honor of being called a body. A mat of bacteria or a group of skin cells is a very different thing from an array of cells that we would call an individual. This is an essential distinction; a thought experiment will help us see the difference.

What happens if you take away some bacteria from a mat of bacteria? You end up with a smaller mat of bacteria. What happens when you remove some cells of a human or fish, say from the heart or brain? You could end up with a dead human or fish, depending on which cells you remove.

So the thought experiment reveals one of the defining features of bodies: our component parts work together to make a greater whole. But not all parts of bodies are equal; some parts are absolutely required for life. Moreover, in bodies, there is a division of labor between parts; brains, hearts, and stomachs have distinct functions. This division of labor extends to the smallest levels of structure, including the cells, genes, and proteins that make bodies.

The body of a worm or a person has an identity that the constituent parts—organs, tissues, and cells—lack. Our skin cells, for example, are continually dividing, dying, and being sloughed off. Yet you are the same individual you were seven years ago, even though virtually every one of your skin cells is now different: the ones you had back then are dead and gone, replaced by new ones. The same is true of virtually every cell in our bodies. Like a river that remains the same despite changes in its course, water content, even size, we remain the same individuals despite the continual turnover of our parts.

And despite this continual change, each of our organs "knows" its size and place in the body. We grow in the correct proportions because the growth of the bones in our arms is coordinated with the growth of the bones in our fingers and our skulls. Our skin is smooth because cells can communicate to maintain its integrity and the regularity of its surface. Until something out of the ordinary happens, like, for instance, we get a wart. The cells inside the wart aren't following the rules: they do not know when to stop growing.

When the finely tuned balance among the different parts of bodies breaks down, the individual creature can die. A cancerous tumor, for example, is born when one batch of cells no longer cooperates with others. By dividing endlessly, or by failing to die properly, these cells can destroy the necessary balance that makes a living individual person. Cancers break the rules that allow cells to cooperate with one another. Like bullies who break down highly cooperative societies, cancers behave in their own best interest until they kill their larger community, the human body.

What made all this complexity possible? For our distant ancestors to go from single-celled creatures to bodied ones, as they did over a billion years ago, their cells had to utilize new mechanisms to work together. They needed to be able to communicate with one another. They needed to be able to stick together in new ways. And they needed to be able to make new things, such as the molecules that make our organs distinct. These features—the glue between cells, the ways cells can "talk" to each other, and the molecules that cells make—constitute the toolkit needed to build all the different bodies we see on earth.

The invention of these tools amounted to a revolution. The shift from single-celled animals to animals with bodies reveals a whole new world. New creatures with whole new capabilities came about: they got big, they moved around, and they developed new organs that helped them sense, eat, and digest their world.

DIGGING UP BODIES

Here's a humbling thought for all of us worms, fish, and humans: most of life's history is the story of single-celled creatures. Virtually everything we have talked about thus far—animals with hands, heads, sense organs, even body plans—has been around for only a small fraction of the earth's history. Those of us who teach paleontology often use the analogy of the "earth year" to illustrate how tiny that fraction is. Take the entire 4.5-billion-year history of the earth and scale it down to a single year, with January 1 being the origin of the earth and midnight on December 31 being the present. Until June, the only organisms were single-celled microbes, such as algae, bacteria, and amoebae. The first animal with a head did not appear until October. The first human appears on December 31. We, like all the animals and plants that have ever lived, are recent crashers at the party of life on earth.

The vastness of this time scale becomes abundantly clear when we look at the rocks in the world. Rocks older than 600 million years are generally devoid of animals or plants. In them we find only single-celled creatures or colonies of algae. These colonies form mats or strands; some colonies are doorknob-shaped. In no way are these to be confused with bodies.

The first people to see the earliest bodies in the fossil record had no idea what they were looking at. Between 1920 and 1960 really odd fossils started popping up from all around the world. In the 1920s and 1930s, Martin Gurich, a German paleontologist working in what is today Namibia, discovered a variety of impressions of what looked like animal bodies. Shaped like disks and plates, these things seemed unremarkable: they could have been primitive algae or jellyfish living in ancient seas.

In 1947, an Australian mining geologist named Reginald Sprigg happened upon a locality where the undersides of the rocks contained impressions of disks, ribbons, and fronds. Working around an abandoned mine in the Ediacara Hills of South Australia, Sprigg uncovered a collection of these fossils and described them dutifully. Over time, similar impressions became known from every continent of the world except Antarctica. Sprigg's creatures seemed strange, but few people really cared about them.

The reason for the collective paleontological yawn was that these fossils were thought to come from the relatively young rocks of the Cambrian era, when many animal fossils with primitive bodies were already known. Sprigg's and Gurich's fossils sat relatively unnoticed, an assemblage of not terribly exciting, if weird, impressions from a period already well represented in the museum collections of the world.

In the mid-1960s, Martin Glaessner, a charismatic Austrian ex-pat living in Australia, changed all that. After comparing these

A timescale for events in the history of life. Notice the extremely long period of time during which there were no bodies on earth, only single-celled organisms living alone or in colonies.

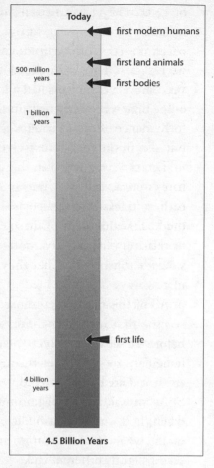

Today

first modern humans

first land animals

500 million years — first bodies

1 billion years

first life

4 billion years

4.5 Billion Years

rocks to those in other parts of the world, Glaessner showed that without a doubt these fossils were 15 million to 20 million years older than originally thought. They were no dull collection of impressions—rather, Gurich, Sprigg, and others were seeing the earliest bodies.

These fossils came from the period known as the Precambrian, a period thought to be devoid of life. Our understanding of the antiquity of life had just exploded. Paleontological curiosities became scientific jewels.

The Precambrian disks, ribbons, and fronds are clearly the oldest creatures with bodies. As we'd expect from other early animal fossils, they include representatives of some of the most primitive animals on the planet today: sponges and jellyfish. Other Precambrian fossils look like nothing known. We can tell that they are impressions of something with a body, but their patterns of blobs, stripes, and shapes match no living creature.

One message from this is very clear: creatures with many cells began to populate the seas of the planet by 600 million years ago. These creatures had well-defined bodies and weren't just colonies of cells. They have patterns of symmetry that, in some cases, resemble those of living forms. As for those that cannot be compared directly with living forms, different parts of their bodies nevertheless have specialized structures. This implies that the Precambrian organisms had a level of biological organization that at the time was utterly new on the planet.

Evidence of these changes is seen not only in the fossil bodies but also in the rocks themselves. With the first bodies come the first trackways. Etched in the rocks are the first signs that creatures were actually crawling and squirming through the ooze. The earliest trackways, small ribbon-shaped scrapes in the ancient mud, show that some of these creatures with bodies were capable of relatively complicated motions. Not only did they have bodies with identifiable parts, but they were actually using them to move in new ways.

All of this makes total sense. We see the first bodies before we see the first body plans. We see the first primitive body plans before we see the first body plans with heads, and so on. Like the imaginary zoo we walked through in the first chapter, the rocks of the world are highly ordered.

As we said at the beginning of this section, we are after the when, how, and why of bodies. The Precambrian discoveries tell us the when. To see the how, and ultimately the why, we need to take a slightly different tack.

OUR OWN BODY OF EVIDENCE

A photo could never capture just how much of our bodies is to be found within those Precambrian disks, fronds, and ribbons. What could we humans, with all our complexity, ever share with impressions in rocks, particularly ones that look like crinkled jellyfish and squashed rolls of film?

The answer is profound and, when we see the evidence, inescapable: the "stuff" that holds us together—that makes our bodies possible—is no different from what formed the bodies of Gurich's and Sprigg's ancient impressions. In fact, the scaffolding of our entire body originated in a surprisingly ancient place: single-celled animals.

What holds a clump of cells together, whether they form a jellyfish or an eyeball? In creatures like us, that biological glue is astoundingly complicated; it not only holds our cells together, but also allows cells to communicate and forms much of our structure. The glue is not one thing; it is a variety of different molecules that connect and lie between our cells. At the microscopic level, it gives each of our tissues and organs its distinctive appearance and function. An eyeball looks different from a leg bone whether we look at it with the naked eye or under a microscope. In fact, much of the difference between a leg bone and an eye rests in the ways the cells and materials are arranged deep inside.

Every fall for the past several years, I have driven medical students crazy with just these concepts. Nervous first-year students must learn to identify organs by looking at random slides of tissue under a microscope. How do they do this?

The task is a little like figuring out what country you are in by looking at a street map of a small village. The task is doable, but we need the right clues. In organs, some of the best clues lie in the shape of cells and how they attach to one another; it is also important to be able to identify the stuff that lies between them. Tissues have all kinds of different cells, which attach to one another in different ways: some regions have strips or columns of cells; in others, cells are randomly scattered and loosely attached to one another. These areas, where cells are loosely packed, are often filled with materials that give each tissue its characteristic physical properties. For instance, the minerals that lie between bone cells determine the hardness of bone, whereas the looser proteins in the whites of our eyes make the wall of the eyeball more pliant.

Our students' ability to identify organs from microscope slides, then, comes from knowing how cells are arranged and what lies between the cells. For us, there is a deeper meaning. The molecules that make these cellular arrangements possible are the molecules that make bodies possible. If there were no way to attach cells to one another, or if there were no materials between cells, there would be no bodies on the earth—just batches of cells. This means that the starting point for understanding how and why bodies arose is to see these molecules: the molecules that help cells stick together, the molecules that allow them to communicate with one another, and the substances that lie between cells.

To understand the relevance of this molecular structure to our bodies, let's focus in detail on one part: our skeleton. Our skeleton is a powerful example of how tiny molecules can have a big impact on the structure of our body and exemplifies general principles that apply to all the body's parts. Without skeletons, we would be formless masses of goo. Living on land would not be easy or even possible. So much of our basic biology and behavior is made possible by our skeleton that we often take it for granted. Every time we walk, play piano, inhale, or chew food we have our skeleton to thank.

A great analogy for the workings of our skeleton is a bridge. The strength of a bridge depends on the sizes, shapes, and proportions of its girders and cables. But also, importantly, the strength of the bridge depends on the microscopic properties of the materials from which it is made. The molecular structure of steel determines how strong it is and how far it will bend before breaking. In the same way, our skeleton's strength is based on the sizes and shapes of our bones, but also on the molecular properties of our bones themselves.

Let's go for a run to see how. As we jog along a path, our muscles contract, our back, arms, and legs move, and our feet push against the ground to move us forward. Our bones and joints function like a giant complex of levers and pulleys that make all that movement possible. Our body's movements are governed by basic physics: our ability to run is in large part based on the size, shape, and proportions of our skeleton and the configuration of our joints. At this level, we look like a big machine. And like a machine, our design matches our functions. A world-class high jumper has different bone proportions from a champion sumo wrestler. The proportions of the legs of a rabbit or a frog, specialized to hop and jump, are different from those of a horse.

Now, let's take a more microscopic view. Pop a slice of a femur under the microscope, and you will immediately see what gives bone its distinctive mechanical properties. The cells are highly organized in places, particularly on the outer rim of the bone. Some cells stick together, whereas others are separated. Between the separated cells are the materials that define the strength of bone. One of them is the rock, or crystal, known as hydroxyapatite, which we discussed in Chapter 4. Hydroxyapatite is hard the way concrete is: strong when compressed, less strong if twisted or bent. So, like a building made of bricks or concrete, bones are shaped so as to maximize their compressive functions and minimize twisting and bending, something Galileo recognized in the seventeenth century.

The other molecule found between our bone cells is the most common protein in the entire human body. If we magnify it 10,000 times with an electron microscope, we see something that looks like a rope consisting of bundles of little molecular fibers. This molecule, collagen, also has the mechanical properties of a rope. Rope is relatively strong when pulled, but it collapses when compressed; think of the two teams in a tug-of-war running toward the middle. Collagen, like rope, is strong when pulled but weak when the ends are pushed together.

Bone is composed of cells that sit in a sea of hydroxyapatite, collagen, and some other, less common molecules. Some cells stick together; other cells float inside these materials. The strength of bone is based on collagen's strength when pulled, and on hydroxyapatite's strength when compressed.

Cartilage, the other tissue in our skeleton, behaves somewhat differently. During our jog, it was the cartilage in our joints that provided the smooth surfaces where our bones glided against one another. Cartilage is a much more pliant tissue than bone; it can bend and smush as forces are applied to it. The smooth operation of the knee joint, as well as most of the other joints we used during our jog, depends on having relatively soft cartilage. When healthy cartilage is compressed it always returns to its native shape, like a kitchen sponge. During each step of our run, our entire body mass slams against the ground at some speed. Without these protective caps at our joints our bones would grind against one another: a very unpleasant and debilitating outcome of arthritis.

The pliability of cartilage is a property of its microscopic structure. The cartilage at our joints has relatively few cells, and these cells are separated by a lot of filling between them. As with bone, it is the properties of this interstitial filling that largely determine the mechanical properties of the cartilage.

Collagen fills much of the space between cartilage cells (as well as the cells of our other tissues). What really gives cartilage its pliancy is another kind of molecule, one of the most extraordinary in the whole body. This kind of molecule, called a proteoglycan complex, gives cartilage strength when squeezed or compressed. Shaped like a giant three-dimensional brush, with a long stem and lots of little branches, the proteoglycan complex is actually visible under a microscope. It has an amazing property relevant to our abilities to walk and move, thanks to the fact that the tiniest branches love to attach to water. A proteoglycan, then, is a molecule that actually swells up with water, filling up until it's like a giant piece of Jell-O. Take this piece of gelatin, wrap collagen ropes in and around it, and you end up with a substance that is both pliant and somewhat resistant to tension. This, essentially, is cartilage. A perfect pad for our joints. The role of the cartilage cells is to secrete these molecules when the animal is growing and maintain them when the animal is not.

The ratios among the various materials define much of the mechanical differences among bone, cartilage, and teeth. Teeth are very hard and, predictably, there is lots of hydroxyapatite and relatively little collagen within the enamel. Bone has relatively more collagen, less hydroxyapatite, and no enamel. Consequently, it is not as hard as teeth. Cartilage has lots of collagen and no hydroxyapatite, and is loaded with proteoglycans. It is the softest of the tissues in our skeleton. One of the main reasons our skeletons look and work as they do is that these molecules are deployed in the right places in the right proportions.

What does all this have to do with the origin of bodies? One property is common to animals, whether they have skeletons or not: all of them, including clumps of cells, have molecules that lie between their cells, specifically different kinds of collagens and proteoglycans. Collagen seems particularly important: the most common protein in animals, it makes up over 90 percent of the body's protein by weight. Bodybuilding in the distant past meant that molecules like these had to be invented.

Something else is essential for bodies: the cells in our bones have to be able to stick together and talk to one another. How do bone cells attach to one another, and how do different parts of bone know to behave differently? Here is where much of our bodybuilding kit lies.

Bone cells, like every cell in our bodies, stick to one another by means of tiny molecular rivets, of which there is a vast diversity. Some bind cells the way contact cement holds the soles of shoes together: one molecule is firmly attached to the outer membrane of one cell, another to the outer membrane of a neighboring cell. Thus attached to both cell membranes, the glue forms a stable bond between the cells.

Other molecular rivets are so precise that they bind selectively, only to the same kind of rivet. This is a hugely significant feature because it helps organize our bodies in a fundamental way. These selective rivets enable cells to organize themselves and ensure that bone cells stick to bone cells, skin to skin, and so on. They can organize our bodies in the absence of other information. If we put a number of cells, each with a different kind of this type of rivet, on a dish and let the cells grow, the cells will organize themselves. Some might form balls, others sheets, as the cells sort out by the numbers and kinds of rivets they have.

But arguably the most important connection between cells lies in the ways that they exchange information with one another. The precise pattern of our skeleton, in fact of our whole body, is possible only because cells know how to behave. Cells need to know when to divide, when to make molecules, and when to die. If, for example, bone or skin cells behaved randomly—if they divided too much or died too little—then we would be very ugly or, worse, very dead.

Cells communicate with one another using "words" written as molecules that move from cell to cell. One cell can "talk" to the next by sending molecules back and forth. For instance, in a relatively simple form of cell-to-cell communication, one cell will emit a signal, in this case a molecule. This molecule will attach to the outer covering, or membrane, of the cell receiving the signal. Once attached to the outer membrane, the molecule will set off a chain reaction of molecular events that travels from the outer membrane all the way, in many cases, to the nucleus of the cell. Remember that the genetic information sits inside the nucleus. Consequently, this molecular signal can cause genes to be turned on and off. The end result of all this is that the cell receiving the information now changes its behavior: it may die, divide, or make new molecules in response to the cue from the other cell.

At the most basic level, these are the things that make bodies possible. All animals with bodies have structural molecules like collagens and proteoglycans, all of them have the array of molecular rivets that hold cells together, and all of them have the molecular tools that allow cells to communicate with one another.

We now have a search image to understand the how of body origins. To see how bodies arose, we need to look for these molecules in the most primitive bodies on the planet, and then, ultimately, in creatures that have no body at all.

BODYBUILDING FOR BLOBS

What does the body of a professor share with a blob? Let's look at some of the most primitive bodies alive today to find the answer.

One of these creatures has the dubious distinction of almost never being seen in the wild. In the late 1880s, a strangely simple creature was discovered living on the glass walls of an aquarium. Unlike anything else alive, it looked like a mass of goo. The only thing we can compare it with is the alien creature in the Steve McQueen movie *The Blob*. Recall that the Blob was an amorphous glop that, after dropping in from outer space, engulfed its prey: dogs, people, and eventually small diners in little towns in Pennsylvania. The Blob's digestive end was on its underside: we never saw it; we only heard the shrieks of creatures caught there. Shrink the Blob down to between 200 and 1,000 cells, about two millimeters in diameter, and we have the enigmatic living creature known as a placozoan. Placozoans have only four types of cells, which make a very simple body shaped like a small plate. It is a real body, though. Some of the cells on the undersurface are specialized for digestion; others have flagella, which beat to move the creature around. We have little idea of what they eat in the wild, where they live, or what their natural habitat is. Yet these simple blobs reveal something terrifically important: with a small number of specialized cells, these primitive creatures already have a division of labor among their parts.

Much of what is interesting about bodies already exists in placozoans. They have true bodies, albeit primitively organized ones. In searching through their DNA and examining the molecules on the surface of their cells, we find that much of our bodybuilding apparatus is already there. Placozoans have versions of the molecular rivets and cell communication tools we see in our own bodies.

Our bodybuilding apparatus is found in blobs simpler than some of Reginald Sprigg's ancient impressions. Can we go further, to even more primitive kinds of bodies? Part of the answer lies in a piece of classic kitchenware: the sponge. At first glance, sponges are unremarkable. The body of a sponge consists of the sponge matrix itself; not a living material, it is a form of silica (glassy material) or calcium carbonate (a hard shell-like material) with some collagen interspersed. Right off the bat, that makes sponges interesting. Recall that collagen is a major part of our intercellular spaces, holding cells and many tissues together. Sponges may not look it, but they already have one of the earmarks of bodies.

In the early 1900s, H.V.P. Wilson showed just how amazing sponges really are. Wilson came to the University of North Carolina as its first professor of biology in 1894. There he went on to train a cadre of American biologists who were to define the field of genetics and cell biology in North America for the next century. As a young man, Wilson decided to focus his life's research on, of all things, sponges. One of his experiments revealed a truly remarkable capability of these apparently simple creatures. He ran them through a kind of sieve, which broke them down to a set of disaggregated cells. Wilson put the now completely disaggregated, amoeba-like cells in a dish and watched them. At first, they crawled around on the surface of the dish. Then, something surprising happened: the cells came together. First, they formed red cloudy balls of cells. Next, they gained more organization, with cells becoming packed in definite patterns. Finally, the clump of cells would form an entire new sponge body, with the various types of cells assuming the appropriate positions. Wilson was watching a body come together almost from scratch. If we were like sponges, then the Steve Buscemi character who gets minced in the woodchipper in the Coen brothers' movie *Fargo* would have been just fine. In fact, he might have been invigorated by the experience, as his cells might have aggregated to form many different versions of him.

It is the cells within sponges that make them useful in understanding the origin of bodies. The inside of the sponge is usually a hollow space that can be divided into compartments, depending on the species. Water flows through the space, directed by a very special kind of cell. These cells are shaped like goblets with the cup part facing the inside of the sponge. Tiny cilia extending from the rim of the goblet beat and capture food particles in the water. Also extending from the goblet part of each of these cells is a large flagellum. The concerted action of the flagella of these little beater cells moves water and food through the pores of the sponge. Other cells on the inside of the sponge process the particles of food. Still others line the outside and can contract when the sponge needs to change its shape as water currents change.

A sponge seems a far cry from a body, yet it has many of the most important properties of bodies: its cells have a division of labor; the cells can communicate with one another; and the array of cells functions as a single individual. A sponge is organized, with different kinds of cells in different places doing different things. It is a far cry from a human body with trillions of precisely packaged cells, but it shares some of the human body's features.

Most significantly, the sponge has much of the cell adhesion, communication, and scaffolding apparatus that we have. Sponges are bodies, albeit very primitive and relatively disorganized ones.

Like placozoans and sponges, we have many cells. Like them, our bodies show a division of labor among parts. The whole molecular apparatus that holds bodies together is also present: the rivets that hold cells together; the various devices that help cells signal to one another; and many of the molecules that lie between cells. Like us and all other animals, placozoans and sponges even have collagen. Unlike us, they have very primitive versions of all these features: instead of twenty-one collagens, sponges have two; whereas we have hundreds of different types of molecular rivets, sponges have a small fraction of that number. Sponges are simpler than we and have fewer kinds of cells, but the basic bodybuilding apparatus is there.

Placozoans and sponges are about as simple as bodies get nowadays. To go any further, we have to search for the things that build our bodies in creatures that have no bodies at all: single-celled microbes.

How do you compare a microbe to an animal with a body? Are the tools that build bodies in animals present in single-celled creatures? If so, and if they are not building bodies, what are they doing?

The most straightforward way to begin to answer these questions involves looking inside the genes of microbes to search for any similarities to animals. The earliest comparisons between animal and microbial genomes revealed a striking fact: in many single-celled animals, much of the molecular machinery for cell adhesion, interaction, and so on is just not there. Some analyses even suggested that more than eight hundred of these kinds of molecules are found only in animals with bodies while they are absent in single-celled creatures. This would seem to support the notion that the genes that help cells unite to make bodies arose together with the origin of bodies. And at first glance, it seems to make sense that the tools to build bodies should arise in lockstep with bodies themselves.

The story turned upside down when Nicole King, of the University of California at Berkeley, studied the organisms called choanoflagellates. King's choice of subject was no accident. From work on DNA, she knew that choanoflagellates are likely the closest microbe relatives of animals with bodies, placozoans, and sponges. She also suspected that hidden in the genes of choanoflagellates are versions of the DNA that make our bodies.

Nicole was aided in her search by the Human Genome Project, an enterprise that has succeeded in mapping all the genes in our bodies. With the success of the Human Genome Project came many other mapping studies: we've had the Rat Genome Project, the Fly Genome Project, the Bumblebee Genome Project—there are even ongoing projects to sequence the genomes of sponges, placozoans, and microbes. These maps are a gold mine of information because they enable us to compare the bodybuilding genes in many different species. They also gave Nicole the genetic tools to study her choanoflagellates.

Choanoflagellates look remarkably like the goblet-shaped cells inside a sponge. In fact, for a long time, many people thought that they were just degenerate sponges—sponges without all the other cells. If this were the case, then the DNA of choanoflagellates should resemble that of a bizarre sponge. It doesn't. When parts of the DNA of choanoflagellates were compared with microbe and sponge DNA, the similarity to microbe DNA turned out to be extraordinary. Choanoflagellates are single-celled microbes.

The genetic distinction between "single-celled microbe" and "animal with body" completely broke down thanks to Nicole's work on choanoflagellates. Most of the genes that are active in choanoflagellates are also active in animals. In fact, many of those genes are part of the machinery that builds bodies. A few exam-

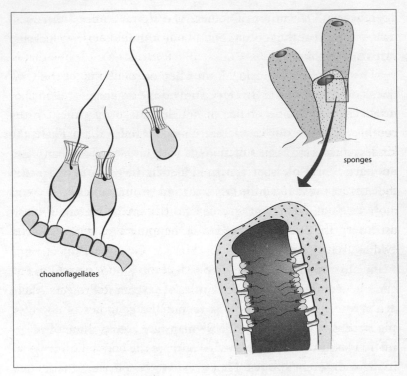

Choanoflagellates (left) and sponges (right).

ples reveal the power of this comparison. Functions of cell adhesion and cell communication, even parts of the molecules that form the matrix between cells and the molecular cascades that ferry a signal from outside the cell to the inside—all are present in choanoflagellates. Collagens are present in choanoflagellates. The various kinds of molecular rivets that hold cells together are also present in choanoflagellates, although they are doing slightly different jobs.

Choanoflagellates even give Nicole a road map for comparing our bodybuilding apparatus to that of other microbes. The fundamental molecular structure that makes collagens and proteoglycan aggregates is known from a number of different kind of microbes. *Streptococcus* bacteria—common in our mouths (and, one hopes, rare in other places)—have on their cell surface a molecule that is very similar to collagen. It has the same molecular signature, but does not aggregate to form ropes or sheets as collagens do in animals. Likewise, some of the sugars that make up proteoglycan complexes inside our cartilage are seen in the walls of different kinds of bacteria. Their functions in both viruses and bacteria are not particularly pleasant. They are associated with the ways that these agents invade and infect cells and, in many cases, become more virulent. Many of the molecules that microbes use to cause us misery are primitive versions of the molecules that make our own bodies possible.

This sets up a puzzle. In the fossil record, we see nothing but microbes for the first 3.5 billion years of earth history. Then, suddenly, over a span of perhaps 40 million years, all kinds of bodies appear: plant bodies, fungal bodies, animal bodies; bodies everywhere. Bodies were a real fad. But, if you take Nicole's work at face value, the potential to build bodies was in place well before bodies ever hit the scene. Why the rush for bodies after such a very long time with no bodies at all?

A PERFECT STORM IN THE ORIGIN OF BODIES

Timing is everything. The best ideas, inventions, and concepts don't always win. How many musicians, inventors, and artists were so far ahead of their time that they flopped and were forgotten, only to be rediscovered later? We need look no further than poor Heron of Alexandria, who, perhaps in the first century A.D., invented the steam turbine. Unfortunately, it was regarded as a toy. The world wasn't ready for it.

The history of life works the same way. There is a moment for everything, perhaps even for bodies. To see this, we need to understand why bodies might have come about in the first place.

One theory about this is extremely simple: Perhaps bodies arose when microbes developed new ways to eat each other or avoid being eaten? Having a body with many cells allows creatures to get big. Getting big is often a very good way to avoid being eaten. Bodies may have arisen as just that kind of defense.

When predators develop new ways of eating, prey develop new ways of avoiding that fate. This interplay may have led to the origin of many of our bodybuilding molecules. Many microbes feed by attaching and engulfing other microbes. The molecules that allow microbes to catch their prey and hold on to them are likely candidates for the molecules that form the rivet attachments between cells in our bodies. Some microbes can actually communicate with each other by making compounds that influence the behavior of other microbes. Predator-prey interactions between microbes often involve molecular cues, either to ward off potential predators or to serve as lures enticing prey to come close. Perhaps signals like these were precursors to the kinds of signals that our own cells use to exchange information to keep our bodies intact.

We could speculate on this ad infinitum, but more exciting would be some tangible experimental evidence that shows how predation could bring about bodies. That is essentially what Martin Boraas and his colleagues provided. They took an alga that is normally single-celled and let it live in the lab for over a thousand generations. Then they introduced a predator: a single-celled creature with a flagellum that engulfs other microbes to ingest them. In less than two hundred generations, the alga responded by becoming a clump of hundreds of cells; over time, the number of cells dropped until there were only eight in each clump. Eight turned out to be the optimum because it made clumps large enough to avoid being eaten but small enough so that each cell could pick up light to survive. The most surprising thing happened when the predator was removed: the algae continued to reproduce and form individuals with eight cells. In short, a simple version of a multicellular form had arisen from a no-body.

If an experiment can produce a simple body-like organization from a no-body in several years, imagine what could happen in billions of years. The question then becomes not how could bodies arise, but why didn't they arise sooner?

Answers to this puzzle might lie in the ancient environment in which bodies arose: the world may not have been ready for bodies.

A body is a very expensive thing to have. There are obvious advantages of becoming a creature with a large body: besides avoiding predators, animals with bodies can eat other, smaller creatures and actively move long distances. Both of these abilities allow the animals to have more control over their environment. But both consume a lot of energy. Bodies require even more energy as they get larger, particularly if they incorporate collagen. Collagen requires a relatively large amount of oxygen for its synthesis and would have greatly increased our ancestors' need for this important metabolic element.

But the problem was this: levels of oxygen on the ancient earth were very low. For billions of years oxygen levels in the atmosphere did not come close to what we have today. Then, roughly a billion years ago, the amount of oxygen increased dramatically and has stayed relatively high ever since. How do we know this? From the chemistry of rocks. Rocks from about a billion years ago show the telltale signature of having been formed with increasing amounts of oxygen. Could the rise in oxygen in the atmosphere be linked to the origin of bodies?

It may have taken the paleontological equivalent of a perfect storm to bring about bodies. For billions of years, microbes developed new ways of interacting with their environment and with one another. In doing so, they hit on a number of the molecular parts and tools to build bodies, though they used them for other purposes. A cause for the origin of bodies was also in place: by a billion years ago, microbes had learned to eat each other. There was a reason to build bodies, and the tools to do so were already there.

Something was missing. That something was enough oxygen on the earth to support bodies. When the earth's oxygen increased, bodies appeared everywhere. Life would never be the same.

CHAPTER EIGHT

MAKING SCENTS

I n the early 1980s, there was tension between molecular biologists and people who worked on whole organisms—ecologists, anatomists, and paleontologists. Anatomists, for example, were seen as quaintly out-of-date, hopelessly entranced by an antiquated kind of science. Molecular biology was revolutionizing our approach to anatomy and developmental biology, so much so that the classical disciplines, such as paleontology, seemed to be dead ends in the history of biology. I was made to feel that, because of my love of fossils, I was going to be replaced by one of those new automated DNA sequencers.

Twenty years later, I'm still digging in the dirt and cracking rocks. I'm also collecting DNA and looking at its role in development. Debates usually begin as either-or scenarios. Over time, all-or-nothing positions give way to a more realistic approach. Fossils and the geological record remain a very powerful source of evidence about the past; nothing else reveals the actual environments and transitional structures that existed during the history of life. As we've seen, DNA is an extraordinarily powerful window into life's history and the formation of bodies and organs. Its role is particularly important where the fossil record is silent. Large parts of bodies—soft tissues, for example—simply do not fossilize readily. In these cases, the DNA record is virtually all we have.

Extracting DNA from bodies is incredibly easy, so easy you can do it in your kitchen. Take a handful of tissue from some plant or animal—peas, or steak, or chicken liver. Add some salt and water and pop everything in a blender to mush up the tissue. Then add some dish soap. Soap breaks up the membranes that surround all the cells in the tissue that were too small for the blender to handle. After that, add some meat tenderizer. The meat tenderizer breaks up some of the proteins that attach to DNA. Now you have a soapy, meat-tenderized soup, with DNA inside. Finally, add some rubbing alcohol to the mix. You'll have two layers of liquid: soapy mush on the bottom, clear alcohol on top. DNA has a real attraction to alcohol and will move into it. If a goopy white ball appears in the alcohol, you've done everything right. That goop is the DNA.

You are now in a position to use that white glop to understand many of the basic connections we have with the rest of life. The trick, on which we spend countless hours and dollars, comes down to comparing DNA's structure and function in different species. Here is the counterintuitive bit. By extracting DNA from *any* tissue, say the liver, of different species, you can actually decipher the history of virtually any part of our body, including our sense of smell. Locked inside that DNA, whether it comes from liver, blood, or muscle, is much of the apparatus we use to detect odors in our environment. Recall that all our cells contain the same DNA; what differs is which bits of DNA are active. The genes involved in the sense of smell are present in all of our cells, although they are active only in the nasal area.

As we all know, odors elicit impulses in our brains that can have a profound impact on the way we perceive our world. A whiff might lead us to recall the schoolrooms of our childhood or the musty coziness of our grandparents' attic, each occasion bringing long-buried feelings to the surface. More essentially, smells can help us to survive. The smell of tasty food gets us hungry; the smell of sewage makes us feel ill. We are hardwired to avoid rotten eggs. Want to sell your home? It would be far better to have bread baking in the oven than cabbage boiling on the stovetop when prospective buyers come by. We collectively invest vast sums in our sense of smell: in 2005 the perfume industry generated $24 billion of business in the United States alone. All of this attests to how deeply embedded our sense of smell is inside of us. It is also very ancient.

Our sense of smell allows us to discriminate among five thousand to ten thousand odors. Some people can detect the odor molecules in a green bell pepper at a concentration of less than one part per trillion. That is like picking out one grain of sand from a mile-long beach. How do we do that?

What we perceive as a smell is our brain's response to a cocktail of molecules floating in the air. The molecules that we ultimately register as an odor are tiny, light enough to be suspended in the air. As we breathe or sniff, we suck these odor molecules into our nostrils. From there, the odor molecules go to an area behind our nose where they are trapped by the mucous lining of our nasal passages. Inside this lining is a patch of tissue containing millions of nerve cells, each with little projections into the mucous membrane. When the molecules in the air bind to the nerve cells, signals are sent to our brain. Our brain records these signals as a smell.

The molecular part of smelling works like a lock-and-key mechanism. The lock is the odor molecule; the key is the receptor on the nerve cell. A molecule captured by the mucous membranes in our nose interacts with a receptor on the nerve cell. Only when the molecule attaches to the receptor is a signal sent to our brain. Each receptor is tuned to a different kind of molecule, so a particular odor might involve lots of molecules and, accordingly, lots of receptors sending signals to our brains.

The best analogy for smell comes from music: a chord. A chord is made up of several notes acting together as one. In the same way, an odor is the product of signals from lots of receptors keyed

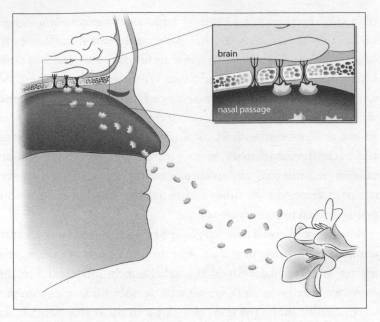

Molecules (enlarged many, many times) from a flower waft through the air. These molecules attach to receptors inside the lining of our nasal cavities. Once the molecules attach, a signal is sent to our brain. Each smell is composed of many different molecules attaching to different receptors. Our brain integrates these signals as we perceive a smell.

to different odor molecules. Our brain perceives these different impulses as one smell.

As in fish, amphibians, reptiles, mammals, and birds, much of our sense of smell is housed inside our skull. Like the other animals, we have one or more holes through which we bring air inside, and then a set of specialized tissues where the chemicals in the air can interact with neurons. We can trace the patterns of these holes, spaces, and membranes from fish to man and find a general pattern. The most primitive living animals with skulls, jawless fish such as lampreys and hagfish, have a single nostril that leads to a sac inside the skull. Water goes into this blind sac, and it is there that smelling takes place; the main difference from us being that lampreys and hagfish extract odors from water instead of air. Our closest fish relatives have an arrangement somewhat like ours: the water enters a nostril and ultimately goes to a cavity linked with the mouth. Fish like lungfish or *Tiktaalik* have two kinds of nostrils: an external one and an internal one. In this, they are a lot like us. Sit with your mouth closed and breathe. Air enters an external nostril and travels through your nasal cavities to enter the back of your throat via internal passageways. Our fish ancestors had internal and external nostrils, too, and to nobody's surprise these are the same fish that have arm bones and other features in common with us.

Our sense of smell contains a deep record of our history as fish, amphibians, and mammals. A major breakthrough in understanding this occurred in 1991 when Linda Buck and Richard Axel discovered the large family of genes that give us our sense of smell.

Buck and Axel used three major assumptions to design their

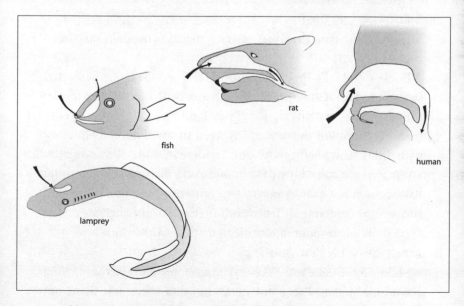

Nasal openings and the flow of odor molecules from jawless fish to man.

experiments. First, they came up with a reasoned hypothesis, based on work done in other laboratories, about what the genes that make odor receptors might have looked like. Experiments showed that odor receptors have a characteristic structure with a number of molecular loops that help them convey information across a cell. This was a big clue, because Buck and Axel could then search the genome of a mouse for every gene that makes this structure. Second, they assumed that the genes for these receptors had to have a very specific activity—they should be active only in the tissues involved with smell. This makes sense: if something is involved in smelling, then it should be restricted to the tissues specialized for that purpose. Third—and this last was a big assumption—Axel and Buck reasoned that there wasn't only one or even a small number of these genes, there had to be lots of them. This hypothesis was based on the fact that many different kinds of chemicals can stimulate different smells. If there was a one-to-one match between each chemical type and a receptor/gene specialized for it, then there had to be many, many genes. But, given the data they had at the time, this needn't have been true.

Buck and Axel's three assumptions were borne out perfectly. They found genes that had the characteristic structure of the receptor they were looking for. They found that all of these genes were active only in the tissues involved in smelling, the olfactory epithelium. And finally, they found a huge number of these genes. It was a home run. Then, Buck and Axel discovered something truly astounding: fully 3 percent of our entire genome is devoted to genes for detecting different odors. Each of these genes makes a receptor for an odor molecule. For this work, Buck and Axel shared the Nobel Prize in 2004.

Following Buck and Axel's success, people started fishing around for olfactory receptor genes in other species. It turns out that such genes are a living record of some major transitions in the history of life. Take the water-to-land transition, over 365 million years ago. There are two kinds of smelling genes: one is specialized for picking up chemical scents in the water, the other specialized for air. The chemical reaction between odor molecule and receptor is different in water and air, hence the need for slightly different receptors. As we'd expect, fish have water-based receptors in their nasal neurons, mammals and reptiles have air-based ones.

This discovery helps us make sense of the state of affairs in the most primitive fish alive on the planet today—the jawless fish such as lampreys and hagfish. It turns out that these creatures have, unlike more advanced fish and mammals, neither "air" nor "water" genes; rather, their receptors combine both types. The implication is clear: these primitive fish arose before the smelling genes split into two types.

Jawless fish reveal another very important point: they have a very small number of odor genes. Bony fish have more, and still more are seen in amphibians and reptiles. The number of odor genes has increased over time, from relatively few in primitive creatures such as jawless fish, to the enormous number seen in mammals. We mammals, with over a thousand of these genes, devote a huge part of our entire genetic apparatus just to smelling. Presumably, the more of these genes an animal has, the more acute its ability to discern different kinds of smells. In this light, our large number of odor genes makes sense—mammals are highly specialized smelling animals. Just think of what effective trackers dogs can be.

But where do all our extra odor genes come from? Did they just pop out of the blue? How this expansion happened seems obvious when we look at the structure of the genes. If you compare the odor genes of a mammal with the handful of odor genes in a jawless fish, the "extra" genes in mammals are all variations on a theme: they look like copies, albeit modified ones, of the genes in jawless fish. This means that our large number of odor genes arose by many rounds of duplication of the small number of genes present in primitive species.

This leads us to a paradox. Humans devote about 3 percent of our genome to odor genes, just like every other mammal. When geneticists looked at the structure of the human genes in more detail, they found a big surprise: fully three hundred of these thousand genes are rendered completely functionless by mutations that have altered their structure beyond repair. (Other mammals do use these genes.) Why have so many odor genes if so many of them are entirely useless?

Dolphins and whales, of all creatures, offer an insight to help us answer this question. Like all mammals, dolphins and whales have hair, breasts, and a three-boned middle ear. Their mammalian history is also recorded in their smelling genes: lacking fish-like water-specialized genes, cetaceans have mammalian air-specialized genes. The mammalian history of whales and dolphins is even written in the DNA of their odor perception apparatus. But there is an interesting puzzle: dolphins and whales no longer use their nasal passages to smell. What are these genes doing? The former nasal passage has been modified into a blowhole, which is used in breathing, not in smelling. This has had a remarkable effect on the smelling genes: all of a cetacean's odor genes are present, but not one is functional.

What has happened to the smell genes of dolphins and whales also happens in many other species' genes. Mutations crop up in genomes from generation to generation. If a mutation knocks out the function of a gene, the result can be dangerous, or even lethal. But what happens if a mutation knocks out the function of a gene that does nothing? There is a lot of mathematical theory that says the obvious: such mutations will just silently get passed on from generation to generation. This is exactly what appears to have happened in dolphins. Their smell genes are no longer needed, given the blowhole, so the mutations that knocked out their function just accumulate over time. The genes have no use, but they remain present in the DNA as silent records of evolution.

But humans do have a sense of smell, so why have so many of our odor genes been knocked out? Yoav Gilad and his colleagues answered this question by comparing genes among different primates. He found that primates that develop color vision tend to have large numbers of knocked-out smell genes. The conclusion is clear. We humans are part of a lineage that has traded smell for sight. We now rely on vision more than on smell, and this is reflected in our genome. In this trade-off, our sense of smell was deemphasized, and many of our olfactory genes became functionless.

We carry a lot of baggage in our noses—or, more precisely, in the DNA that controls our sense of smell. Our hundreds of useless olfactory genes are left over from mammal ancestors who relied more heavily on the sense of smell to survive. In fact, we can take these comparisons deeper still. Like photocopies that lose their fidelity as they are repeatedly copied, our olfactory genes get more dissimilar as we compare ourselves to successively more primitive creatures. Our genes are similar to primates', less similar to other mammals', less similar still to reptiles', amphibians', fishes', and so on. That baggage is a silent witness to our past; inside our noses is a veritable tree of life.

CHAPTER NINE

VISION

Only once in my entire career have I found the eye of a fossil creature. I wasn't in the field on an expedition, I was in the back room of a mineral shop in a small town in northeast China. My colleague Gao Keqin and I were studying the earliest known salamanders, beautiful fossils collected from Chinese rocks about 160 million years old. We had just finished a collecting trip to some sites Gao knew about. The locations were secret, because these salamander fossils have serious monetary value for the farmers who typically find them. What makes them special is that impressions of the soft tissue, such as gills, guts, and the notochord, are often preserved. Private collectors love them because fossils of this quality are exceedingly rare. By the time we ended up at the mineral shop, Gao and I had already collected a number of really beautiful ancient salamanders of our own from his sites.

This particular mineral dealer had gotten his hands on one of the best salamander fossils of all time. Gao wanted us to see it and spent the better part of a day trying to work the deal. The whole visit had a terrifically illicit feel. Gao spent several hours smoking cigarettes with the gentleman, speaking and gesturing in Chinese. Clearly there was some bartering going on, but not knowing Chinese I had no idea what offers were being put on the table. After lots of headshaking and ultimately a big handshake, I was permitted to go to the back room and look at a fossil on the dealer's desk. It was a stunning sight: the body of a larval salamander, no more than three inches long. In it, I could see impressions of the whole animal, all the way down to the little shells it ate as its last meal. And, for the first and only time in my career, I was staring at the eye of an ancient fossil animal.

Eyes rarely make it into the fossil record. As we've seen, the best candidates for preservation as fossils are the hard parts of the animals—bones, teeth, and scales. If we want to understand the history of eyes, then we can use an important fact to our advantage. There is a remarkable diversity of organs and tissues that animals use to capture light, from simple photoreceptor organs in invertebrate animals to the compound eyes of various insects and our own camera-type eye. How do we put this variation to use in understanding how our ability to see developed over time?

The history of our eyes is a lot like that of a car. Take a Chevy Corvette, for example. We can trace the history of the model as a whole—the Corvette—and the history of each of its parts. The 'Vette has a history, beginning with its origins in 1953 and continuing through the different model designs each year. The tires used on the 'Vette also have a history, as does the rubber used in making them. This supplies a great analogy for bodies and organs. Our eyes have a history as organs, but so do eyes' constituent parts, the cells and tissues, and so do the genes that make those parts. Once we identify these multiple layers of history in our organs, we understand that we are simply a mosaic of bits and pieces found in virtually everything else on the planet.

Much of the processing of the images we see actually happens inside our brains: the role of the eye is to capture light in a way that it can be carried to the brain for processing as an image. Our eyes, like those of every creature with a skull and backbone, are like little cameras. After light from the outside enters the eye, it is focused on a screen at the back of the eyeball. Light travels through several layers as it traverses this path. First it passes through the cornea, a thin layer of clear tissue that covers the lens. The amount of light that enters the eye is controlled by a diaphragm, called the iris, which dilates and contracts by the action of involuntary muscles. The light then passes through the lens, which, as a camera does, focuses the image. Tiny muscles surround the lens; as these muscles contract, they change the lens's shape, thus focusing images from near and far. A healthy lens is clear and made up of special proteins that give it its distinctive shape and light-gathering properties. These proteins, known as lens crystallins, are exceptionally long-lived, allowing the lens to continue functioning as we age. The screen on which all of the light is projected, the retina, is loaded with blood vessels and light receptors. These light receptors send signals to our brain that we then interpret as images. The retina absorbs the light via sensitive light-gathering cells. There are two types of such cells: one is very sensitive to light, the other less so. The more sensitive cells record only in black and white; the less sensitive cells record in color. If we look around the animal world, we can assess whether animals are specialized for daylight or night by looking at the percentages of each type of light-sensing cell in their eyes. In humans these cells make up about 70 percent of all the sensory cells in our body. That is a clear statement about how important vision is to us.

Our camera-like eye is common to every creature with a skull, from fish to mammals. In other groups of animals we find different eyes, ranging from simple patches of cells specialized to detect light, to eyes with compound lenses such as those found in flies, to primordial versions of our own eye. The key to understanding the history of our eyes is to understand the relationship

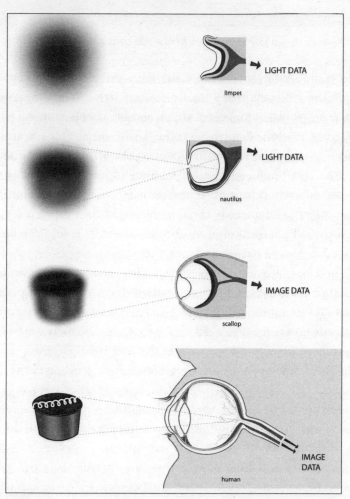

Eyes come into focus: from primitive light-capturing devices in invertebrates to our camera-type eye with a lens. As eyes evolve, visual acuity increases.

between the structures that make our camera-eye and those that make all the other kinds of eyes. To do this, we will study the molecules that gather light, the tissues we use to see, and the genes that make the whole thing.

LIGHT-GATHERING MOLECULES

The really important work in the light-gathering cells happens inside the molecule that actually collects light. When this molecule absorbs light, it changes shape and breaks up into two parts. One part is derived from vitamin A, the other from a protein known as an opsin. When the opsin breaks off, it initiates a chain reaction that leads to a neuron sending an impulse to our brain. We use different opsins to see in black and white and in color. Just as an inkjet printer needs three or four inks to print in color, we need three light-gathering molecules to see in color. For black-and-white vision, we use only one.

These light-gathering molecules change shape in the light, then recharge in the dark and go back to their normal state. The process takes a few minutes. We all know this from personal experience: go from a bright place into a dark room and it is virtually impossible to see faint objects. The reason is that the light-gathering molecules need time to recharge. After a few minutes, vision in the dark returns.

Despite the stunning variety of photoreceptor organs, every animal uses the same kind of light-capturing molecule to do this job. Insects, humans, clams, and scallops all use opsins. Not only can we trace the history of eyes through differences in the structure of their opsins, but we have good evidence that we can thank bacteria for these molecules in the first place.

Essentially, an opsin is a kind of molecule that conveys information from the outside of a cell to the inside. To pull off this feat, it needs to carry a chemical across the membrane that encircles a cell. Opsins use a specialized kind of conductor that takes a series of bends and loops as it travels from the outside to the inside of the cell. But this twisted path the receptor takes through the cell membrane is not random—it has a characteristic signature.

Where else is this twisted path seen? It is identical to parts of certain molecules in bacteria. The very precise molecular similarities in this molecule suggest a very ancient property of all animals extending all the way to our shared history with bacteria. In a sense, modified bits of ancient bacteria lie inside our retinas, helping us to see.

We can even trace some major events in the history of our eyes by examining opsins in different animals. Take one of the major events in our primate past, the development of rich color vision. Recall that humans and our closest ape relatives, the Old World monkeys, have a very detailed kind of color vision that relies on three different kinds of light receptors. Each of these receptors is tuned to a different kind of light. Most other mammals have only two kinds of receptors and so cannot discriminate as many colors as we can. It turns out that we can trace the origin of our color vision by looking at the genes that make the receptors. The two kinds of receptors most mammals have are made by two kinds of genes. Of our three receptor-making genes, two are remarkably like one of those in other mammals. This seems to imply that our color vision began when one of the genes in other mammals duplicated and the copies specialized over time for different light sources. As you'll remember, a similar thing happened with odor receptor genes.

This shift may be related to changes in the flora of the earth millions of years ago. It helps to think what color vision was likely good for when it first appeared. Monkeys that live in trees would benefit because color vision enabled them to discriminate better among many kinds of fruits and leaves and select the most nutritious among them. From studying the other primates that have color vision, we can estimate that our kind of color vision arose about 55 million years ago. At this time we find fossil evidence of changes in the composition of ancient forests. Before this time, the forests were rich in figs and palms, which are tasty but all of the same general color. Later forests had more of a diversity of plants, likely with different colors. It seems a good bet that the switch to color vision correlates with a switch from a monochromatic forest to one with a richer palette of colors in food.

TISSUES

Animal eyes come in two flavors; one is seen in invertebrates, the other in vertebrates, such as fish and humans. The central idea is that there are two different ways of increasing the light-gathering surface area in eye tissue. Invertebrates, such as flies and worms, accomplish this by having numerous folds in the tissue, while our lineage expands the surface area by having lots of little projections extending from the tissue like tiny bristles. A host of other differences also relate to these different kinds of designs. Lacking fossils at the relevant phase of history, it would seem that we would never be able to bridge the differences between our eyes and those of invertebrates. That is, until 2001, when Detlev Arendt thought to study the eyes of a very primitive little worm.

Polychaetes are among the most primitive living worms known. They have a very simple segmented body plan, and they also have two kinds of light-sensing organs: an eye and, buried under their skin, a part of their nervous system that is specialized to pick up light. Arendt took these worms apart both physically and genetically. Knowledge of the gene sequence of our opsin genes and the structure of our light-gathering neurons gave Arendt the tools to study how polychaetes are made. He found that they had elements of both kinds of animal photoreceptors. The normal "eye" was made up of neurons and opsins like the eye of any invertebrate. The tiny photoreceptors under the skin were another matter altogether. They had "vertebrate" opsins and cellular structure even with the little bristle-like projections, but in primitive form.

Arendt had found a living bridge, an animal with both kinds of eyes, one of which—our kind—existed in a very primitive form. When we look to primitive invertebrates, we find that the different kinds of animal eyes share common parts.

GENES

Arendt's discovery leads to yet another question. It is one thing for eyes to share common parts, but how can eyes that look so different—such as those of worms, flies, and mice—be closely related? For the answer, let us consider the genetic recipe that builds eyes.

At the turn of the twentieth century, Mildred Hoge was recording mutations in fruit flies when she found a fly that had no eyes whatsoever. This mutant was not an isolated case, and Hoge discovered that she could breed a whole line of such flies, which she named *eyeless*. Later, a similar mutation was discovered in mice. Some individuals had small eyes; others lacked whole portions of the head and face, including their eyes. A similar condition in humans is known as aniridia; affected individuals are missing large pieces of their eyes. In these very different creatures—flies, mice, and humans—geneticists were finding similar kinds of mutants.

A breakthrough came in the early 1990s, when laboratories applied new molecular techniques to understand how eyeless mutants affected eye development. Mapping the genes, they were able to localize the bits of DNA responsible for the mutations. When the DNA was sequenced, it turned out that the fly, mouse, and human genes responsible for eyelessness had similar DNA structures and sequences. In a very real sense, they are the same gene.

What did we learn from this? Scientists had identified a single gene that, when mutated, produced creatures with small eyes or no eyes at all. This meant that the normal version of the gene was a major trigger for the formation of eyes. Now came the chance to do experiments to ask a whole other kind of question. What happens when we mess with the gene, turning it on and off in the wrong places?

Flies were an ideal subject for this work. During the 1980s, a number of very powerful genetic tools were developed through work on flies. If you knew a gene, or a DNA sequence, you could make a fly lacking the gene or, the reverse, a fly with the gene active in the wrong places.

Using these tools, Walter Gehring started playing around with the *eyeless* gene. Gehring's team was able to make the *eyeless* DNA active pretty much anywhere they wanted: in the antenna, on the legs, on the wings. When his team did this, they found something stunning. If they turned on the *eyeless* gene in the antenna, an eye grew there. If they turned on the *eyeless* gene on a body segment, an eye developed there. Everywhere they turned on the gene, they would get a new eye. To top it all off, some of the misplaced eyes showed a nascent ability to respond to light. Gehring had uncovered a major trigger in the formation of eyes.

Gehring didn't stop there; he began swapping genes between species. They took the mouse equivalent of *eyeless*, *Pax 6*, and turned it on in a fly. The mouse gene produced a new eye. And not just any eye—a fly eye. Gehring's lab found they could use the *mouse* gene to trigger the formation of an extra *fly* eye anywhere: on the back, on a wing, near the mouth. What Gehring had found was a master switch for eye development that was virtually the same in a mouse and a fly. This gene, *Pax 6*, initiated a complex chain reaction of gene activity that ultimately led to a new fly eye.

We now know that *eyeless*, or *Pax 6*, controls development in everything that has eyes. The eyes may look different—some with a lens, some without; some compound, some simple—but the genetic switches that make them are the same.

When you look into eyes, forget about romance, creation, and the windows into the soul. With their molecules, genes, and tissues derived from microbes, jellyfish, worms, and flies, you see an entire menagerie.

CHAPTER TEN

EARS

The first time you see the inside of the ear is a letdown: the real machinery is hidden deep inside the skull, encased in a wall of bone. Once you open the skull and remove the brain, you need to chip with a chisel to remove that wall. If you are really good, or very lucky, you'll make the right stroke and see it—the inner ear. It resembles the kind of tiny coiled snail shell you find in the dirt in your lawn.

The ear may not look like much, but it is a wonderful Rube Goldberg contraption. When we hear, sound waves are funneled into the outside flap, the external ear. The sound waves enter the ear and make the eardrum rattle. The eardrum is attached to three little bones, which shake along with it. One of these ear bones is attached to the snail-shell structure by a kind of plunger. The shaking of the ear bone causes the plunger to go up and down. This causes some fluid inside the snail shell to move around. Swishing fluid bends hairs on sensory cells that trigger a nerve impulse to the brain, which interprets this as sound. Next time you are at a concert, just imagine all the stuff flying around in your head.

This structure allows us to distinguish three parts to the ear: external, middle, and inner. The external ear is the visible part. The middle contains the little ear bones. Finally, the inner ear consists of the sensory cells, the fluid, and the tissues that surround them. These three components of ears enable us to structure our discussion in a very convenient way.

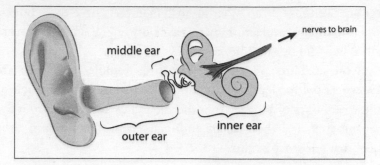

Of the three parts of our ear—the outer, middle, and inner—the inner ear is the most ancient and the part that controls the nerve impulses sent to the brain.

The part of the ear we can see, the flap on which we hang our glasses, is a relatively new evolutionary addition to bodies. Confirm this on your next trip to the aquarium or zoo. How many sharks, bony fish, amphibians, and reptiles have external ears? The pinna—the flap of the external ear—is found only in mammals. Some amphibians and reptiles have visible external ears, but they have no pinna. Often the external ear is only a membrane that looks like the top of a drum.

The elegance of our connection to sharks and bony fish is revealed when we look inside our ears. Ears might seem an unlikely place for a human-shark connection, especially since sharks don't have ears. But the connection is there. Let's start with the ear bones.

THE MIDDLE EAR—THE THREE EAR BONES

Mammals are very special. With hair and milk-producing glands, we can easily be distinguished from other creatures. It surprises most people to learn that some of the most distinctive traits of mammals lie inside the ear. The bones of the mammalian middle ear are like those of no other animal: mammals have three bones, whereas reptiles and amphibians have only one. Fish have none at all. Where did our middle ear bones come from?

Some anatomy: recall that our three middle ear bones are known as the malleus, incus, and stapes. As we've seen, each of these ear bones is derived from the gill arches: the stapes from the second arch, and the malleus and incus from the first arch. It is here that our story begins.

In 1837, the German anatomist Karl Reichert was looking at embryos of mammals and reptiles to understand how the skull forms. He followed the gill arches of different species to understand where they ended up in the various skulls. As he did this again and again, he found something that appeared not to make any sense: two of the ear bones in the mammals corresponded to pieces of the jaw in the reptiles. Reichert could not believe his eyes, and his monograph reveals his excitement. As he describes the ear-jaw comparison, his prose departs from the normally staid description of nineteenth-century anatomy to express shock, even wonderment, at this discovery. The conclusion was inescapable: the same gill arch that formed part of the jaw of a reptile formed ear bones in mammals. Reichert proposed a notion that even he could barely believe—that parts of the ears of mammals are the same thing as parts of the jaws of reptiles. Things get more difficult when we realize that Reichert proposed this several decades before Darwin propounded his notion of a family tree for life. What does it mean to call structures in two different species "the same" without a notion of evolution?

Much later, in 1910 and 1912, the German anatomist Ernst Gaupp picked up on Reichert's work and published an exhaustive study on the embryology of mammalian ears. Gaupp provided more detail and, given the times, interpreted Reichert's work in an evolutionary framework. Gaupp's story went like this: the three middle ear bones reveal the tie between reptiles and mammals.

The single bone in the reptilian middle ear is the same as the stapes of mammals; both are second-arch derivatives. The explosive bit of information, though, was that the two other middle ear bones of mammals—the malleus and the incus—evolved from bones set in the back of the reptilian jaw. If this was indeed the case, then the fossil record should show bones shifting from the jaw to the ear during the origin of mammals. The problem was that Gaupp worked only on living creatures and didn't fully appreciate the role that fossils could play in his theory.

Beginning in the 1840s a number of new kinds of fossil creatures were becoming known from discoveries in South Africa and Russia. Often abundantly preserved, whole skeletons of dog-size animals were being unearthed. As they were discovered, many of them were crated and shipped to Richard Owen in London for identification and analysis. Owen was struck that these creatures had a mélange of features. Parts of their skeleton looked reptile-like. Other parts, notably their teeth, looked like mammals. And these were not isolated finds. It turns out that these "mammal-like reptiles" were the most common skeletons being uncovered at many fossil sites. Not only were they very common, there were many kinds. In the years after Owen, these mammal-like reptiles became known from other parts of the world and from several different time periods in earth history. They formed a beautiful transitional series in the fossil record between reptile and mammal.

Until 1913, embryologists and paleontologists were working in isolation from one another. At this time, the American paleontologist W. K. Gregory, of the American Museum of Natural History, saw an important link between Gaupp's embryos and the African fossils. The most reptilian of the mammal-like reptiles had only a single bone in its middle ear; like other reptiles, it had a jaw composed of many bones. Something remarkable was revealed as Gregory looked at the successively more mammalian mammal-like reptiles, something that would have floored Reichert had he been alive: a continuum of forms showing beyond doubt that over time the bones at the back of the reptilian jaw got smaller and smaller, until they ultimately lay in the middle ear of mammals. The malleus and incus did indeed evolve from jawbones. What Reichert and Gaupp observed in embryos was buried in the fossil record all along, just waiting to be discovered.

Why would mammals need a three-boned middle ear? This little linkage forms a lever system that allows mammals to hear higher-frequency sounds than animals with a single middle ear bone. The origin of mammals involved not only new patterns of chewing, as we saw in Chapter 4, but new ways of hearing. In fact, this shift was accomplished not by evolving new bones per se, but by repurposing existing ones. Bones originally used by reptiles to chew evolved in mammals to assist in hearing.

So much for the malleus and incus. Where, though, does the stapes come from?

If I simply showed you an adult human and a shark, you would never guess that this tiny bone deep inside a human's ear is the same thing as a large rod in the upper jaw of a fish. Yet, developmentally, these bones are the same thing. The stapes is a second-arch bone, as is the corresponding bone in a shark and a fish—the hyomandibula. But the hyomandibula is not an ear bone; recall that fish and sharks do not have ears. In our aquatic cousins, this bone is a large rod that connects the upper jaw to the braincase. Despite the apparent differences in the function and shape of these bones, the similarities between the hyomandibula and the stapes extend even to the nerves that supply them. The key nerve for the functioning of both bones is the second-arch nerve, the facial nerve. We thus have a situation where two very different bones have similar developmental origins and patterns of innervation. Is there an explanation for this?

Again, we look to the fossils. As we trace the hyomandibula from sharks to creatures like *Tiktaalik* to amphibians, we can see

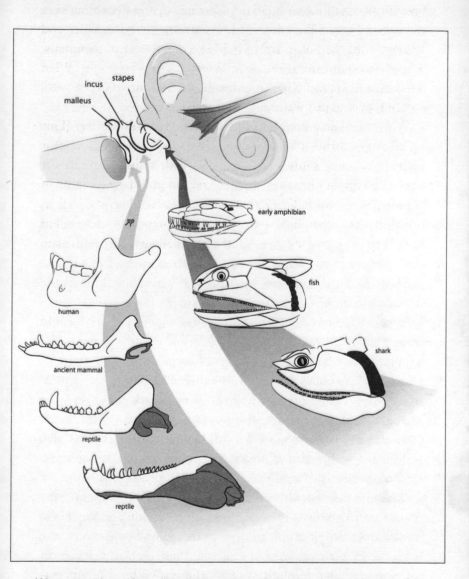

We can trace bones from gill arches to our ears, first during the transition from fish to amphibian (right), and later during the shift from reptile to mammal (left).

how it gets smaller and smaller, ultimately shifting position from the upper jaw to play a role in hearing. The name changes, too. When it is big and supporting the jaw, we call it a hyomandibula; when it is small and functions in hearing, it is known as a stapes. This shift happened when the descendants of fish began to walk on land. Hearing in water is different from hearing on land, and the small size and position of the stapes makes it ideal for picking up vibrations in air. The new ability came about by modifying the upper jawbone of a fish.

Our middle ear contains a record of two of the great transformations in the history of life. The origin of our stapes, and its transformation from a jaw support bone to an ear bone, began when fish started to walk on land. The other big event took place during the origin of mammals, when bones at the back of a reptile jaw became our malleus and incus.

Now let's go further inside the ear—to the inner ear.

THE INNER EAR—GELS MOVING AND HAIRS BENDING

Move through the external ear, go deeper inside, past the eardrum and three middle ear bones, and you end up deep inside the skull. Here you will find the inner ear—tubes and some gel-filled sacs. In humans, as in other mammals, the bony tubes take the snail-shell shape that is so strikingly apparent in the anatomy lab.

The inner ear has different parts dedicated to different functions. One part is used in hearing, another in telling us which way our head is tilted, and still another in recording how fast our head is accelerating or stopping. In carrying out each of these functions, the inner ear works in roughly the same way.

The several parts of the inner ear are filled with a gel that can move. Specialized cells send hairlike projections into this gel. When the gel moves, the hairs on the ends of the these cells bend. When these hairs bend, the nerve cells send an electrical impulse to the brain, where it is recorded as sound, position, or acceleration.

To envision the structure that tells us where our head is in space, imagine a Statue of Liberty snow globe. The snow globe is made of plastic and filled with gel. When you shake it, the gel moves and the "snow" falls on Lady Liberty. Now imagine a snow globe made of a flexible membrane. Pick it up and tilt it, and the whole thing will flop about, causing the gel inside to swish around. This, on a much smaller scale, is what we have inside our ears. When we bend our heads, these contraptions flop around,

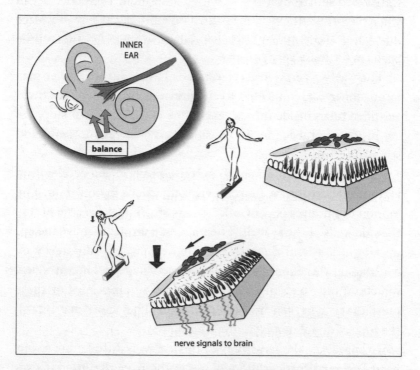

Each time you tilt your head, the tiny rocks on the fluid-filled sacs move. In doing so, they bend nerve endings inside the sacs and cause an impulse to be sent to your brain saying "Your head is tilted."

causing the usual chain of events: the gel inside swishes, the hair projections on the nerves bend, and an impulse is sent back to our brains.

In us, this whole system is made even more sensitive by the presence of tiny rock-like structures on top of the membrane. As we bend our heads, the rocks accentuate the flopping of the membrane, causing the gel to move even more. This increases the sensitivity of the system, enabling us to perceive small differences in position. Tilt your head, and little rocks inside your skull move.

You can probably imagine how tough it would be to live in outer space. Our sensors are tuned to work in the earth's gravity, not in a gravity-free space capsule. Floating around, our eyes recording one version of up and down, our inner ear sensors totally confused, it is all too easy to get sick. Space sickness has been a real problem for these very reasons.

The way we perceive acceleration is based on yet another part of our inner ear, connected to the previous two. There are three gel-filled tubes inside the ear; each time we accelerate or stop, the gel inside the tubes moves, causing the nerve cells to bend and stimulate a current.

The whole system we use to perceive position and acceleration is connected to our eye muscles via connections in our brain. The motion of our eyes is controlled by six small muscles attached to the side walls of the eyeball. The muscles contract to move the eye up, down, left, and right. We can move our eyes voluntarily by contracting these muscles each time we decide to look in a new direction; but some of the most fascinating properties of these muscles relate to their involuntary action. They move our eyes all the time, without our even thinking about it.

To appreciate the sensitivity of this eye-muscle link, move your head back and forth while looking at the page. Keep your eyes fixed in one place as you move your head.

What happened during this experiment? Your eyes stayed fixed

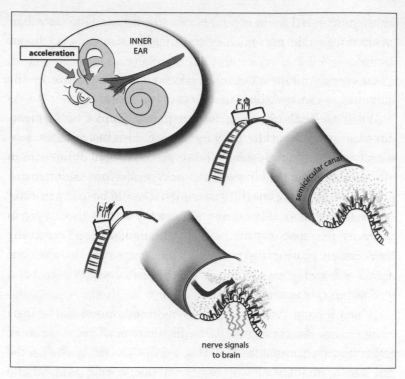

Every time we accelerate, fluid in the inner ear swishes. The swish is transformed into a nerve impulse that is sent to the brain.

on a single point while your head moved. This motion is so commonplace that we take it for granted, but it is incredibly complex. Each of the eight muscles in both eyes is responding to the movement of the head. Sensors in your head, which I'll describe in the next section, record the direction and velocity of your head's movement. These signals are carried to the brain, which then sends out signals telling your eye muscles to fire. Think about that the next time you fix your gaze as your head is moving. This system can misfire, and misfires have much to tell us about our general well-being.

An easy way to understand the inner ear–eye connection is to interfere with it. One way humans do this is to imbibe too much alcohol. Drinking too much ethanol leads us to do silly things because our inhibitions are lowered. Drinking *way* too much gives us the spins. And the spins often predict a lousy morning ahead, hungover, with more spins, nausea, and headache.

When we drink too much, we are putting lots of ethanol into our bloodstream, but the fluid inside our ear tubes initially contains very little. As time passes, however, the alcohol diffuses from our blood into the fluid of the inner ear. Alcohol is lighter than the fluid, so the result of the diffusion is that the fluid in our inner ear becomes less dense. This change in density wreaks havoc on the intemperate among us. Our hair cells are stimulated and our brain thinks we are moving. But we are not moving; we are slumped in a corner or hunched on a barstool. Our brain has been tricked.

The problem extends to our eyes. Our brain thinks we are spinning, and it passes this information to our eye muscles. The eyes twitch in one direction (usually to the right) as we try to track an object moving from side to side. If you prop open the eyes of someone who is stone drunk, you might see this stereotypical twitch, called nystagmus. Police know this well, and often look for nystagmus in people whom they have stopped for driving erratically.

Massive hangovers involve a slightly different response. The day after the binge, your liver has done a remarkably efficient job of removing the alcohol from your bloodstream. Too efficient, for we still have alcohol in the tubes in our ears. That alcohol then diffuses from the gel back into the bloodstream, and in doing so it once more sets the gel in motion: the spins again. Take the same heavy drinker whose eyes you saw twitch to the right the night before and look at him during the hangover. His eyes might still twitch, but in the opposite direction.

We can thank our shared history with sharks and fish for this. If you have ever tried to catch a trout, then you have come up against an organ that is likely an antecedent to our inner ear. As every fisherman knows, trout hold only in certain parts of a stream, typically spots where they can get the best meal while avoiding predators. Often such places are in the shade and in the eddies of the stream's current. Great places for big fish to hold are behind big rocks or fallen logs. Trout, like all fish, have a mechanism that allows them to sense the current and the motion of the water around them, almost like a sense of touch.

Within the skin and bones of the fish, arranged in lines that run the length of the body and head, are small organs with sensory receptors. These receptors lie in small bundles from which they send small hair-like projections into a jelly-filled sac called a neuromast. It helps to think of the snow globe Statue of Liberty again. A neuromast organ is like a tiny one of these, with nerves projecting inside. When the water flows around the fish, it deforms this small sac, thereby bending the hair-like projections of the nerve. Much like the whole system in our ears, this apparatus then sends a signal back to the brain and gives the fish a sense of what the water is doing around them. Sharks and fish can discern the direction in which the water is flowing, and some sharks can even detect distortions of the water, such as are produced by other fish swimming near them. We used a version of this system when we moved our head with a fixed gaze, and we saw it go awry when we propped open the eyes of the inebriated individual at the start of this section. If the ancestor we have in common with sharks and fish had used some other kind of inner ear gel, say one that does not swirl when alcohol is added, we would never spin when drunk.

If you think of our inner ears and neuromast organs as versions of the same thing, you would not be far off. Both come from the same sort of tissue during development, and they share a similar structure. But which came first: neuromasts or inner ears? Here the evidence gets sketchy. If you look at some of the earliest fossils with heads, creatures about 500 million years old, you'll find little pits in their external armor that suggest they had neuromast organs. Unfortunately, we do not know much about the inner ears of these creatures because the preservation of that area of the head is wanting. Until more evidence rolls in, we are left with one of two alternatives: either our inner ears arose from neuromast organs or the other way around. Both scenarios, at their core, reflect a principle we've seen at work in other parts of the body. Organs can come about for one function, only to be repurposed over time for any number of new uses.

In our own ears, there occurred an expansion of the inner ear. The part of our inner ear devoted to hearing is, as in other mammals, huge and coiled. More primitive creatures, such as amphibians and reptiles, have a simple uncoiled inner ear. Clearly, our mammalian forebears obtained a new and better type of hearing. The same is true for the structures that perceive acceleration. We have three canals to record acceleration because we perceive space in three dimensions. The earliest known fish with these canals, a kind of jawless fish like a hagfish, has only one. Then, in other primitive fish, we see two. Finally, most modern fish, and other vertebrates, have three, like us.

We have seen that our inner ear has a history that can be traced to the earliest fish. Remarkably, the neurons inside the gel of our ears have an even more ancient history.

These neurons, called hair cells, have special features that are seen in no other neuron. With fine hair-like projections, consisting of one long "hair" and a series of smaller ones, these neurons lie with a fixed orientation in our inner ear and in a fish's neuromast organ. Recently, people have searched for these cells in other creatures, and have found them not only in animals that do not have sense organs like ours at all but also in animals that have no heads. They are seen in creatures like *Amphioxus*, which we met

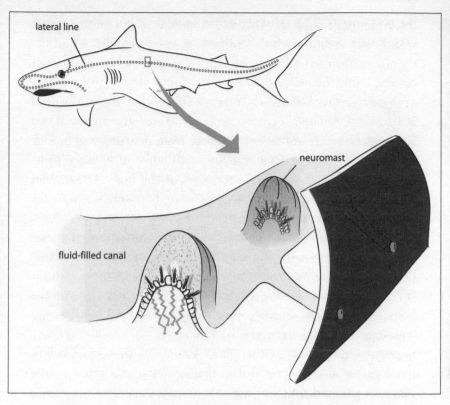

A primitive version of part of our inner ear is embedded in the skin of fish. Small sacs—the neuromasts—are distributed around the body. When they bend, they give the fish information about how the flow of water is changing.

in Chapter 5, that have no ears, eyes, heads, or skulls. Hair cells, then, were around doing other things before our sense organs even hit the scene.

All this is recorded in our genes, of course. If humans or mice have a mutation that knocks out a gene called *Pax 2*, the inner ear fails to form properly. *Pax 2* is active in the ear region and appears to start a chain reaction of gene activity that leads to the development of the inner ear. Go fishing for this gene in more primitive animals and we find *Pax 2* active in the head and, lo and behold, in the neuromasts. The spinning drunk and the fish's water-sensing organs have common genes: evidence of a common history.

JELLYFISH AND THE ORIGINS OF EYES AND EARS

Just like *Pax 6*, which we discussed earlier in connection with eyes, *Pax 2* in ears is a major gene, essential for proper development. Interestingly, a link between *Pax 2* and *Pax 6* suggests that ears and eyes might have had a very ancient common history.

This is where the box jellyfish enters our story. Well known to swimmers in Australia because they have particularly poisonous venom, these jellyfish are different from most others in that they have eyes, more than twenty of them. Most of these eyes are simple pits spread over the jellyfish's epidermis. Other eyes on the body are strikingly similar to our own, with a kind of cornea, a lens, even a nervous structure like ours.

Jellyfish do not have either *Pax 6* or *Pax 2:* they arose before those genes hit the scene. But in the box jellyfish's genes we see something remarkable. The gene that forms the eyes is not *Pax 6*, as we'd expect, but a sort of mosaic that has the structure of *both Pax 6* and *Pax 2*. In other words, this gene looks like a primitive version of other animals' *Pax 6* and *Pax 2*.

The major genes that control our eye and ear correspond to a single gene in more primitive creatures, such as jellyfish. You're probably thinking, So what? The ancient connection between ear and eye genes helps to make sense of things we see in hospital clinics today: a number of human birth defects affect *both* the eyes and the inner ear. All this is a reflection of our deep connections to primitive creatures like the stinging box jellyfish.

CHAPTER ELEVEN

THE MEANING OF IT ALL

THE ZOO IN YOU

My professional introduction to academia happened in the early 1980s, during my college years, when I volunteered at the American Museum of Natural History in New York City. Aside from the excitement of working behind the scenes in the collections of the museum, one of the most memorable experiences was attending their raucous weekly seminars. Each week a speaker would come to present some esoteric study on natural history. Following the presentation, often a fairly low-key affair, the listeners would pick the talk apart point by point. It was merciless. On occasion, the whole thing felt like a human barbecue, with the invited speaker as the spit-roasted main course. Frequently, these debates would devolve into shouting sessions with all the high dudgeon and operatic pantomime of an old silent movie, complete with shaken fists and stomped feet.

Here I was, in the hallowed halls of academe, listening to seminars on taxonomy. You know, taxonomy—the science of naming species and organizing them into the classification scheme that we all memorized in introductory biology. I could not imagine a topic less relevant to everyday life, let alone one less likely to lead eminent senior scientists into apoplexy and the loss of much of their human dignity. The injunction "Get a life" could not have seemed more apt.

The irony is that I now see why they got so worked up. I didn't appreciate it at the time, but they were debating one of the most important concepts in all of biology. It may not seem earth-shattering, but this concept lies at the root of how we compare different creatures—a human with a fish, or a fish with a worm, or anything with anything else. It has led us to develop techniques that allow us to trace our family lineages, identify criminals by means of DNA evidence, understand how the AIDS virus became dangerous, and even track the spread of flu viruses throughout the world. The concept I'm about to discuss supplies the underpinning for much of the logic of this book. Once we grasp it, we see the meaning of the fish, worms, and bacteria that lie inside of us.

The articulation of truly great ideas, of the laws of nature, begins with simple premises that all of us see every day. From simple beginnings, ideas like these extend to explain the really big stuff, like the movement of the stars or the workings of time. In that spirit, I can share with you one true law that all of us can agree upon. This law is so profound that most of us take it completely for granted. Yet it is the starting point for almost everything we do in paleontology, developmental biology, and genetics.

This biological "law of everything" is that every living thing on the planet had parents.

Every person you've ever known has biological parents, as does every bird, salamander, or shark you have ever seen. Technology may change this, thanks to cloning or some yet-to-be-invented method, but so far the law holds. To put it in a more precise form: every living thing sprang from some parental genetic information. This formulation defines parenthood in a way that gets to the actual biological mechanism of heredity and allows us to apply it to creatures like bacteria that do not reproduce the way we do.

The extension of this law is where its power comes in. Here it is, in all its beauty: all of us are modified descendants of our parents or parental genetic information. I'm descended from my mother and father, but I'm not identical to them. My parents are modified descendants of their parents. And so on. This pattern of descent with modification defines our family lineage. It does this so well that we can reconstruct your family lineage just by taking blood samples of individuals.

Imagine that you are standing in a room full of people whom you have never seen before. You are given a simple task: find out how closely related each person in the room is to you. How do you tell who are your distant cousins, your super-distant cousins, your great-granduncles seventy-five times removed?

To answer this question, we need a biological mechanism to guide our thinking and give us a way to test the accuracy of our hypothesized family tree. This mechanism comes from thinking about our law of biology. Knowing how descent with modification works is key to unlocking biological history, because descent with modification can leave a signature, which we can detect.

Let's take a hypothetical humorless, quite unclown-like couple who have children. One of their sons was born with a genetic mutation that gave him a red rubber nose that squeaks. This son grows up and marries a lucky woman. He passes his mutated nose gene to his children, and they all have his red rubber nose that squeaks. Now, suppose one of his offspring gets a mutation that causes him to have huge floppy feet. When this mutation passes to the next generation, all of his children are like him: they have a red rubber nose that squeaks and huge floppy feet. Go one generation further. Imagine that one of these kids, the original couple's great-grandchild, has another mutation: orange curly hair. When this mutation passes to the *next* generation, all of his children will have orange curly hair, a rubber nose that squeaks, and giant floppy feet. When you ask "Who is this bozo?" you'll be inquiring about each of our poor couple's great-great-grandchildren.

This example illustrates a very serious point. Descent with

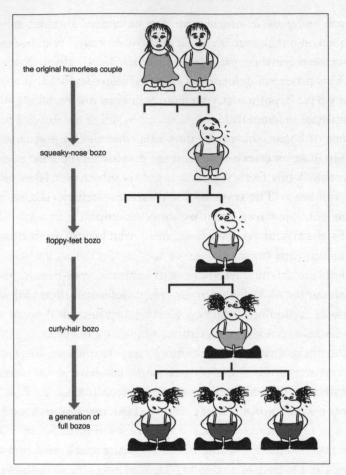

the original humorless couple

squeaky-nose bozo

floppy-feet bozo

curly-hair bozo

a generation of
full bozos

The bozo family tree.

modification can build a family tree, or lineage, that we can identify by characters. It has a signature that we immediately recognize. Like a nested set of Russian dolls, our hypothetical lineage
formed groups within groups, which we recognize by their unique
features. The group of "full bozo" great-great-grandchildren is
descended from an individual who had only the squeaky nose and
the huge floppy feet. This individual was in a group of "proto-

bozos," who are descended from an individual who had only the rubber nose that squeaks. This "pre-proto-bozo" was descended from the original couple, who didn't look overtly clown-like.

This pattern of descent with modification means that you could easily have hypothesized the bozo family tree without me telling you anything about it. If you had a room full of the various generations of bozos, you would have seen that all clown kin are in a group that possesses a squeaky nose. A subset of these have orange hair and floppy feet. Nested within this subset is another group, the full bozos. The key is that the features—orange hair, squeaky nose, big floppy feet—enable you to recognize the groups. These features are your evidence for the different groups, or in this case generations, of clowns.

Replace this family circus with real features—genetic mutations and the body changes that they encode—and you have a lineage that can be identified by biological features. If descent with modification works this way, then our family trees have a signature in their basic structure. So powerful is this truth that it can help us reconstruct family trees from genetic data alone, as we see from the number of genealogical projects currently under way. Obviously, the real world is more complex than our simple hypothetical example. Reconstructing family trees can be difficult if traits arise many different times in a family, if the relationship between a trait and the genes that cause it is not direct, or if traits do not have a genetic basis and arise as the result of changes in diet or other environmental conditions. The good news is that the pattern of descent with modification can often be identified in the face of these complications, almost like filtering out noise from a radio signal.

But where do our lineages stop? Did the bozos stop at the humorless couple? Does my lineage stop at the first Shubins? That seems awfully arbitrary. Does it stop at Ukranian Jews, or northern Italians? How about at the first humans? Or does it continue to 3.8-billion-year-old pond scum, and beyond? Everybody agrees that their own lineage goes back to some point in time, but just how far back is the issue.

If our lineage goes all the way back to pond scum, and does so while following our law of biology, then we should be able to marshal evidence and make specific predictions. Rather than being a random assortment of creatures, all life on earth should show the same signature of descent with modification that we saw among the bozos. In fact, the structure of the entire geological record shouldn't be random, either. Recent additions should appear in relatively young rock layers. Just as I am a more recent arrival than my grandfather in my family tree, so the structure of the family tree of life should also have its parallels in time.

To see how biologists actually reconstruct our relatedness to other creatures, we need to leave the circus and return to the zoo we visited in the first chapter of the book.

A (LONGER) WALK THROUGH THE ZOO

As we've seen, our bodies are not put together at random. Here, I use the word "random" in a very specific sense; I mean that the structure of our bodies is definitely not random with respect to the other animals that walk, fly, swim, or crawl across this earth. Some animals share part of our structure; others do not. There is order to what we share with the rest of the world. We have two ears, two eyes, one head, a pair of arms, and a pair of legs. We do not have seven legs or two heads. Nor do we have wheels.

A walk in the zoo immediately shows our connections to the rest of life. In fact, it will show that we can group much of life in the same way we did with the bozos. Let's go to just three exhibits at first. Start with the polar bears. You can make a long list of the features that you share with polar bears: hair, mammary glands, four limbs, a neck, and two eyes, among lots of other things. Next, consider the turtle across the way. There are definitely similarities, but the list is a bit shorter. You share four limbs, a neck, and two eyes (among other things) with the turtle. But unlike polar bears and you, turtles don't have hair or mammary glands. As for the turtle's shell, that seems unique to the turtle, just as the white fur was unique to the polar bear. Now visit the African fish exhibit. Its inhabitants are still similar to you, but the list of commonalities is even shorter than the list for turtles. Like you, fish have two eyes. Like you, they have four appendages, but those appendages look like fins, not arms and legs. Fish lack, among many other features, the hair and mammary glands that you share with polar bears.

This is beginning to sound like the Russian doll set of groups, subgroups, and sub-subgroups that appeared in the bozo example. Fish, turtles, polar bears, and humans all share some features—heads, two eyes, two ears, and so on. Turtles, polar bears, and humans have all these features, and they also have necks and limbs, features not seen in fish. Polar bears and humans form an even more elite group, whose members have all of these features and also hair and mammary glands.

The bozo example gives us the means to make sense of our walk through the zoo. In the bozos, the pattern of groups reflected descent with modification. The implication is that the full-bozo kids shared a more recent relative than they do with the kids who have only a squeaky nose. That makes sense: the parent of the squeaky-nosed kids is the great-great grandparent of the full bozos. Applying this same approach to the groups we encountered during our zoo walk means that humans and polar bears should share a more recent ancestor than they do with turtles. This prediction is true: the earliest mammal is much more recent than the earliest reptile.

The central issue here is deciphering the family tree of species. Or, in more precise biological terms, their pattern of relatedness.

This pattern even gives us the means to interpret a fossil such as *Tiktaalik* in light of our walk through the zoo. *Tiktaalik* is a wonderful intermediate between fish and their land-living descendants, but the odds of it being our exact ancestor are very remote. It is more like a cousin of our ancestor. No sane paleontologist would ever claim that he or she had discovered "The Ancestor." Think about it this way: What is the chance that while walking through any random cemetery on our planet I would discover an actual ancestor of mine? Diminishingly small. What I would discover is that all of the people buried in these cemeteries—no matter whether that cemetery is in China, Botswana, or Italy—are related to me to different degrees. I can find this out by looking at their DNA with many of the forensic techniques in use in crime labs today. I'd see that some of the denizens of the cemeteries are distantly related to me, others are related more closely. This tree would be a very powerful window into my past and my family history. It would also have a practical application because I could use this tree to understand my predilection to get certain diseases and other facts of my biology. The same is true when we infer relationships among species.

The real power of this family tree lies in the predictions it allows us to make. Chief among these is that as we identify more shared characteristics, they should be consistent with the framework. That is, as I identify features from cells, DNA, and all the other structures, tissues, and molecules in the bodies of these animals, they should support the groupings that we identified during our walk. Conversely, we can falsify our groupings by finding features inconsistent with them. That is, if there exist many traits shared by fish and people that aren't seen in polar bears, our framework is flawed and needs to be revised or jettisoned. In cases where the evidence is ambiguous, we apply a number of statistical tools to assess the quality of the characteristics supporting the arrangements in the family tree. In instances where there is ambiguity, the genealogical arrangement is treated as a working hypothesis until we can find something conclusive to allow us to either accept or reject it.

Some groupings are so strong that, for all intents and purposes, we consider them fact. The fish–turtle–polar bear–human grouping, for example, is supported by characteristics from hundreds of genes and virtually all features of the anatomy, physiology, and cellular biology of these animals. Our fish-to-human framework is so strongly supported that we no longer try to marshal evidence for it—doing so would be like dropping a ball fifty times to test the theory of gravity. The same holds for our biological example. You would have the same chance of seeing your ball go up the fifty-first time you dropped it as you would of finding strong evidence against these relationships.

We can now return to the opening challenge of the book. How can we confidently reconstruct the relationships among long-dead animals and the bodies and genes of recent ones? We look for the signature of descent with modification, we add characteristics, we evaluate the quality of the evidence, and we assess the degree to which our groups are represented in the fossil record. The amazing thing is that we now have tools to probe this hierarchy, using computers and large DNA sequencing labs to perform the same analyses you performed during your walk through the zoo. We now have access to new fossil sites around the world. We can see our bodies' place in the natural world better than we ever could.

From Chapter 1 through Chapter 10, we have shown that deep similarities exist between creatures living today and those long deceased—ancient worms, living sponges, and various kinds of fish. Now, armed with knowledge of the pattern of descent with modification, we can begin to make sense of it all. Enough fun at the circus and zoo. It's time to get down to business.

———

We have seen that inside our bodies are connections to a menagerie of other creatures. Some parts resemble parts of jellyfish, others parts of worms, still others parts of fish. These aren't haphazard similarities. Some parts of us are seen in every other animal; others are very unique to us. It is deeply beautiful to see that there is an order in all these features. Hundreds of characters from DNA, innumerable anatomical and developmental features—all follow the same logic as the bozos we saw earlier.

Let's consider some of the features we've already talked about in the book and show you how they are ordered.

With every other animal on the planet, we share a body composed of *many cells*. Call this group multicellular life. We share the trait of multicellularity with everything from sponges to placozoans to jellyfish to chimpanzees.

A subset of these multicellular animals have *a body plan like ours*, with a front and a back, a top and a bottom, and a left and a right. Taxonomists call this group Bilateria (meaning "bilaterally symmetrical animals"). It includes every animal from insects to humans.

A subset of multicellular animals that have a body plan like ours, with a front and a back, a top and a bottom, and a left and a right, also have *skulls and backbones*. Call these creatures vertebrates.

A subset of the multicellular animals that have a body plan like ours, with a front and a back, a top and a bottom, and a left and a right, and that have skulls, also have *hands and feet*. Call these vertebrates tetrapods (animals with four limbs).

A subset of the multicellular animals that have a body plan like ours, with a front and a back, a top and a bottom, and a left and a right, that have skulls, and that have hands and feet, also have a *three-boned middle ear*. Call these tetrapods mammals.

A subset of the multicellular animals that have a body plan like

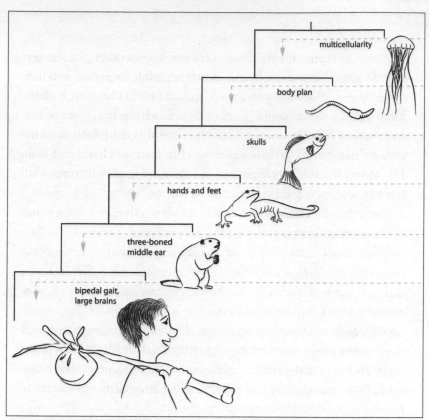

A human family tree, all the way back to jellyfish. It has the same structure as the one for the bozos.

ours with a front and a back, a top and a bottom, and a left and a right, that have skulls and backbones, that have hands and feet, and that have a three-boned middle ear, also have *a bipedal gait and enormous brains.* Call these mammals people.

The power of these groupings is seen in the evidence on which they are based. Hundreds of genetic, embryological, and anatomical features support them. This arrangement allows us to look inside ourselves in an important way.

This exercise is almost like peeling an onion, exposing layer after layer of history. First we see features we share with all other mammals. Then, as we look deeper, we find the features we share with fish. Deeper still are those we share with worms. And so on. Recalling the logic of the bozos, this means that we see a pattern of descent with modification deeply etched inside our own bodies. That pattern is reflected in the geological record. The oldest many-celled fossil is over 600 million years old. The earliest fossil with a three-boned middle ear is less than 200 million years old. The oldest fossil with a bipedal gait is around 4 million years old. Are all these facts just coincidence, or do they reflect a law of biology we can see at work around us every day?

Carl Sagan once famously said that looking at the stars is like looking back in time. The stars' light began the journey to our eyes eons ago, long before our world was formed. I like to think that looking at humans is much like peering at the stars. If you know how to look, our body becomes a time capsule that, when opened, tells of critical moments in the history of our planet and of a distant past in ancient oceans, streams, and forests. Changes in the ancient atmosphere are reflected in the molecules that allow our cells to cooperate to make bodies. The environment of ancient streams shaped the basic anatomy of our limbs. Our color vision and sense of smell has been molded by life in ancient forests and plains. And the list goes on. This history is our inheritance, one that affects our lives today and will do so in the future.

WHY HISTORY MAKES US SICK

My knee was swollen to the size a grapefruit, and one of my colleagues from the surgery department was twisting and bending it to determine whether I had strained or ripped one of the ligaments or cartilage pads inside. This, and the MRI scan that followed, revealed a torn meniscus, the probable result of twenty-five years spent carrying a backpack over rocks, boulders, and scree in the field. Hurt your knee and you will almost certainly injure one or more of three structures: the medial meniscus, the medial collateral ligament, or the anterior cruciate ligament. So regular are injuries to these three parts of your knee that these three structures are known among doctors as the "Unhappy Triad." They are clear evidence of the pitfalls of having an inner fish. Fish do not walk on two legs.

Our humanity comes at a cost. For the exceptional combination of things we do—talk, think, grasp, and walk on two legs—we pay a price. This is an inevitable result of the tree of life inside us.

Imagine trying to jury-rig a Volkswagen Beetle to travel at speeds of 150 miles per hour. In 1933, Adolf Hitler commissioned Dr. Ferdinand Porsche to develop a cheap car that could get 40 miles per gallon of gas and provide a reliable form of transportation for the average German family. The result was the VW Beetle. This history, Hitler's plan, places constraints on the ways we can modify the Beetle today; the engineering can be tweaked only so far before major problems arise and the car reaches its limit.

In many ways, we humans are the fish equivalent of a hot-rod Beetle. Take the body plan of a fish, dress it up to be a mammal, then tweak and twist that mammal until it walks on two legs, talks, thinks, and has superfine control of its fingers—and you have a recipe for problems. We can dress up a fish only so much without paying a price. In a perfectly designed world—one with no history—we would not have to suffer everything from hemorrhoids to cancer.

Nowhere is this history more visible than in the detours, twists, and turns of our arteries, nerves, and veins. Follow some nerves and you'll find that they make strange loops around other organs, apparently going in one direction only to twist and end up in an unexpected place. The detours are fascinating products of our past that, as we'll see, often create problems for us—hiccups and hernias, for example. And this is only one way our past comes back to plague us.

Our deep history was spent, at different times, in ancient oceans, small streams, and savannahs, not office buildings, ski slopes, and tennis courts. We were not designed to live past the age of eighty, sit on our keisters for ten hours a day, and eat Hostess Twinkies, nor were we designed to play football. This disconnect between our past and our human present means that our bodies fall apart in certain predictable ways.

Virtually every illness we suffer has some historical component. The examples that follow reflect how different branches of the tree of life inside us—from ancient humans, to amphibians and fish, and finally to microbes—come back to pester us today. Each of these examples show that we were not designed rationally, but are products of a convoluted history.

OUR HUNTER-GATHERER PAST:
OBESITY, HEART DISEASE, AND HEMORRHOIDS

During our history as fish we were active predators in ancient oceans and streams. During our more recent past as amphibians, reptiles, and mammals, we were active creatures preying on everything from reptiles to insects. Even more recently, as primates, we were active tree-living animals, feeding on fruits and leaves. Early humans were active hunter-gatherers and, ultimately, agriculturalists. Did you notice a theme here? That common thread is the word "active."

The bad news is that most of us spend a large portion of our day being anything but active. I am sitting on my behind at this very minute typing this book, and a number of you are doing the same reading it (except for the virtuous among us who are reading it in the gym). Our history from fish to early human in no way prepared us for this new regimen. This collision between present and past has its signature in many of the ailments of modern life.

What are the leading causes of death in humans? Four of the top ten causes—heart disease, diabetes, obesity, and stroke—have some sort of genetic basis and, likely, a historical one. Much of the difficulty is almost certainly due to our having a body built for an active animal but the lifestyle of a spud.

In 1962, the anthropologist James Neel addressed this notion from the perspective of our diet. Formulating what became known as the "thrifty genotype" hypothesis, Neel suggested that our human ancestors were adapted for a boom-bust existence. As hunter-gatherers, early humans would have experienced periods of bounty, when prey was common and hunting successful. These periods of plenty would be punctuated by times of scarcity, when our ancestors had considerably less to eat.

Neel hypothesized that this cycle of feast and famine had a signature in our genes and in our illnesses. Essentially, he proposed that our ancestors' bodies allowed them to save resources during times of plenty so as to use them during periods of famine. In this context, fat storage becomes very useful. The energy in the food we eat is apportioned so that some supports our activities going on now, and some is stored, for example in fat, to be used later. This apportionment works well in a boom-bust world, but it fails miserably in an environment where rich foods are available 24/7. Obesity and its associated maladies—age-related diabetes, high blood pressure, and heart disease—become the natural state of affairs. The thrifty genotype hypothesis also might explain why we love fatty foods. They are high-value in terms of how much energy they contain, something that would have conferred a distinct advantage in our distant past.

Our sedentary lifestyle affects us in other ways, because our circulatory system originally appeared in more active animals.

Our heart pumps blood, which is carried to our organs via arteries and returned to the heart by way of veins. Because arteries are closer to the pump, the blood pressure in them is much higher than in veins. This can be a particular problem for the blood that needs to return to our heart from our feet. Blood from the feet needs to go uphill, so to speak, up the veins of our legs to our chest. If the blood is under low pressure, it may not climb all the way. Consequently, we have two features that help the blood move up. The first are little valves that permit the blood to move up but stop it from going down. The other feature is our leg muscles. When we walk we contract them, and this contraction serves to pump the blood up our leg veins. The one-way valves and the leg-muscle pumps enable our blood to climb from feet to chest.

This system works superbly in an active animal, which uses its legs to walk, run, and jump. It does not work well in a more sedentary creature. If the legs are not used much, the muscles will not pump the blood up the veins. Problems can develop if blood pools in the veins, because that pooling can cause the valves to fail. This is exactly what happens with varicose veins. As the valves fail, blood pools in the veins. The veins get bigger and bigger, swelling and taking tortuous paths in our legs.

Needless to say, the arrangement of veins can also be a real pain in the behind. Truck drivers and others who sit for long stretches of time are particularly prone to hemorrhoids, another cost of our sedentary lives. During their long hours of sitting, blood pools in the veins and spaces around the rectum. As the blood pools, hemorrhoids form—an unpleasant reminder that we were not built to sit for too long, particularly not on soft surfaces.

PRIMATE PAST: TALK IS NOT CHEAP

Talking comes at a steep price: choking and sleep apnea are high on the list of problems we have to live with in order to be able to talk.

We produce speech sounds by controlling motions of the tongue, the larynx, and the back of the throat. All of these are relatively simple modifications to the basic design of a mammal or a reptile. As we saw in Chapter 5, the human larynx is made up mostly of gill arch cartilages, corresponding to the gill bars of a shark or fish. The back of the throat, extending from the last molar tooth to just above the voice box, has flexible walls that can open and close. We make speech sounds by moving our tongue, by changing the shape of our mouth, and by contracting a number of muscles that control the rigidity of this wall.

Sleep apnea is a potentially dangerous trade-off for the ability to talk. During sleep, the muscles of our throat relax. In most people, this does not present a problem, but in some the passage can collapse so that relatively long stretches pass without a breath. This, of course, can be very dangerous, particularly in people who have heart conditions. The flexibility of our throat, so useful in our ability to talk, makes us susceptible to a form of sleep apnea that results from obstruction of the airway.

Another trade-off of this design is choking. Our mouth leads both to the trachea, through which we breathe, and to our esophagus, so we use the same passage to swallow, breathe, and talk. These three functions can be at odds, for example when a piece of food gets lodged in the trachea.

FISH AND TADPOLE PAST: HICCUPS

This annoyance has its roots in the history we share with fish and tadpoles.

If there is any consolation for getting hiccups, it is that our misery is shared with many other mammals. Cats can be stimulated to hiccup by sending an electrical impulse to a small patch of tissue in their brain stem. This area of the brain stem is thought to be the center that controls the complicated reflex that we call a hiccup.

The hiccup reflex is a stereotyped twitch involving a number of muscles in our body wall, diaphragm, neck, and throat. A spasm in one or two of the major nerves that control breathing causes these muscles to contract. This results in a very sharp inspiration of air. Then, about 35 milliseconds later, a flap of tissue in the back of our throat (the glottis) closes the top of our airway. The fast inhalation followed by a brief closure of the tube produces the "hic."

The problem is that we rarely experience only a single hic. Stop the hiccups in the first five to ten hics, and you have a decent chance of ending the bout altogether. Miss that window, and the bout of hiccups can persist for an average of about sixty hics. Inhaling carbon dioxide (by breathing into the classic paper bag) and stretching the body wall (taking a big inhalation and holding it) can end hiccups early in some of us. But not all. Some cases of pathological hiccups can be extremely prolonged. The longest uninterrupted hiccups in a person lasted from 1922 to 1990.

Our tendency to develop hiccups is another influence of our past. There are two issues to think about. The first is what causes the spasm of nerves that initiates the hiccup. The second is what controls that distinctive hic, the abrupt inhalation–glottis closure. The nerve spasm is a product of our fish history, while the hic is an outcome of the history we share with animals such as tadpoles.

First, fish. Our brain can control our breathing without any conscious effort on our part. Most of the work takes place in the brain stem, at the boundary between the brain and the spinal cord. The brain stem sends nerve impulses to our main breathing muscles. Breathing happens in a pattern. Muscles of the chest, diaphragm, and throat contract in a well-defined order. Consequently, this part of the brain stem is known as a "central pattern generator." This region can produce rhythmic patterns of nerve and, consequently, muscle activation. A number of such generators in our brain and spinal cord control other rhythmic behaviors, such as swallowing and walking.

The problem is that the brain stem originally controlled breathing in fish; it has been jury-rigged to work in mammals. Sharks and bony fish all have a portion of the brain stem that controls the rhythmic firing of muscles in the throat and around the gills. The nerves that control these areas all originate in a well-defined portion of the brain stem. We can even see this nerve arrangement in some of the most primitive fish in the fossil record. Ancient ostracoderms, from rocks over 400 million years old, preserve casts of the brain and cranial nerves. Just as in living fish, the nerves that control breathing extend from the brain stem.

This works well in fish, but it is a lousy arrangement for mammals. In fish, the nerves that control breathing do not have to travel very far from the brain stem. The gills and throat generally surround this area of the brain. We mammals have a different problem. Our breathing is controlled by muscles in the wall of our chest and by the diaphragm, the sheet of muscle that separates our chest from our abdomen. Contraction of the diaphragm controls inspiration. The nerves that control the diaphragm exit our brain just as they do in fish, and they leave from the brain stem, near our neck. These nerves, the vagus and the phrenic nerve, extend from the base of the skull and travel through the chest cavity to reach the diaphragm and the portions of the chest that control breathing.

This convoluted path creates problems; a rational design would have the nerves traveling not from the neck but from nearer the diaphragm. Unfortunately, anything that interferes with one of these nerves can block their function or cause a spasm.

If the odd course of our nerves is a product of our fishy past, the hiccup itself is likely the product of our history as amphibians. Hiccups are unique among our breathing behaviors in that an abrupt intake of air is followed by a closure of the glottis. Hiccups seem to be controlled by a central pattern generator in the brain stem: stimulate this region with an electrical impulse, and we stimulate hiccups. It makes sense that hiccups are controlled by a central pattern generator, since, as in other rhythmic behaviors, a set sequence of events happens during a hic.

It turns out that the pattern generator responsible for hiccups is virtually identical to one in amphibians. And not in just any amphibians—in tadpoles, which use both lungs and gills to breathe. Tadpoles use this pattern generator when they breathe with gills. In that circumstance, they want to pump water into their mouth and throat and across the gills, but they do not want the water to enter their lungs. To prevent it from doing so, they close the glottis, the flap that closes off the breathing tube. And to close the glottis, tadpoles have a central pattern generator in their brain stem so that an inspiration is followed immediately by a closing glottis. They can breathe with their gills thanks to an extended form of hiccup.

The parallels between our hiccups and gill breathing in tadpoles are so extensive that many have proposed that the two phenomena are one and the same. Gill breathing in tadpoles can be blocked by carbon dioxide, just like our hiccups. We can also block gill breathing by stretching the wall of the chest, just as we can stop hiccups by inhaling deeply and holding our breath. Perhaps we could even block gill breathing in tadpoles by having them drink a glass of water upside down.

SHARK PAST: HERNIAS

Our propensity for hernias, at least for those hernias near the groin, results from taking a fish body and morphing it into a mammal.

Fish have gonads that extend toward their chest, approaching their heart. Mammals don't, and therein lies the problem. It is a very good thing that our gonads are not deep in our chest and near our heart (although it might make reciting the Pledge of Allegiance a different experience). If our gonads were in our chest, we wouldn't be able to reproduce.

Slit the belly of a shark from mouth to tail. The first thing you'll see is liver, a lot of it. The liver of a shark is gigantic. Some zoologists believe that a large liver contributes to the buoyancy of the shark. Move the liver away and you'll find the gonads extending up near the heart, in the "chest" area. This arrangement is typical of most fish: the gonads lie toward the front of the body.

In us, as in most mammals, this arrangement would be a disaster. Males continuously produce sperm throughout our lives. Sperm are finicky little cells that need exactly the right range of temperatures to develop correctly for the three months they live. Too hot, and sperm are malformed; too cold, and they die. Male mammals have a neat little device for controlling the temperature of the sperm-making apparatus: the scrotum. As we all know, the male gonads sit in a sac. Inside the skin of the sac are muscles that can expand and contract as the temperature changes. Muscles also lie in our sperm cords. Hence, the cold shower effect: the scrotum will tuck close to the body when it is cold. The whole package rises and falls with temperature. This is all a way to optimize the production of healthy sperm.

The dangling scrotum also serves as a sexual signal in many mammals. Between the physiological advantages of having gonads outside the body wall, and the occasional benefits this provides in

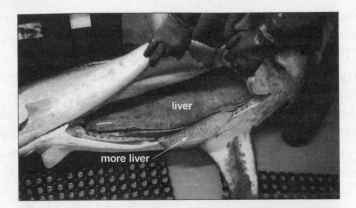

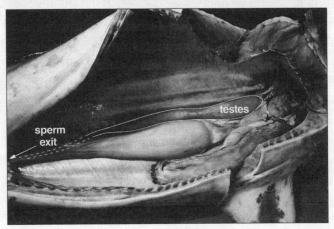

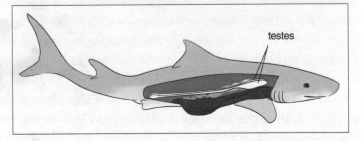

Open a shark and you find a huge liver (top). Push the liver aside and you see gonads, which extend relatively close to the heart, as they do in other primitive creatures. Photos courtesy of Dr. Steven Campana, Canadian Shark Research Laboratory.

securing mates, there are ample advantages for our distant mammalian ancestors in having a scrotum.

The problem with this arrangement is that the plumbing that carries sperm to the penis is circuitous. Sperm travel from the testes in the scrotum through the sperm cord. The cord leaves the scrotum, travels up toward the waist, loops over the pelvis, then goes through the pelvis to travel through the penis and out. Along this complex path, the sperm gain seminal fluids from a number of glands that connect to the tube.

The reason for this absurd route lies in our developmental and evolutionary history. Our gonads begin their development in much the same place as a shark's: up near our livers. As they grow and develop, our gonads descend. In females, the ovaries descend from the midsection to lie near the uterus and fallopian tubes. This ensures that the egg does not have far to travel to be fertilized. In males, the descent goes farther.

The descent of the gonads, particularly in males, creates a weak spot in the body wall. To envision what happens when the testes and spermatic cord descend to form a scrotum, imagine pushing your fist against a rubber sheet. In this example, your fist becomes equivalent to the testes and your arm to the spermatic cord. The problem is that you have created a weak space where your arm sits. Where once the rubber sheet was a simple wall, you've now made another space, between your arm and the rubber sheet, where things can slip. This is essentially what happens in many types of inguinal hernias in men. Some of these inguinal hernias are congenital—when a piece of the gut travels with the testes as it descends. Another kind of inguinal hernia is acquired. When we contract our abdominal muscles, our guts push against the body wall. A weakness in the body wall means that guts can escape the body cavity and be squeezed to lie next to the spermatic cord.

Females are far tougher than males, particularly in this part of the body. Because females do not have a giant tube running

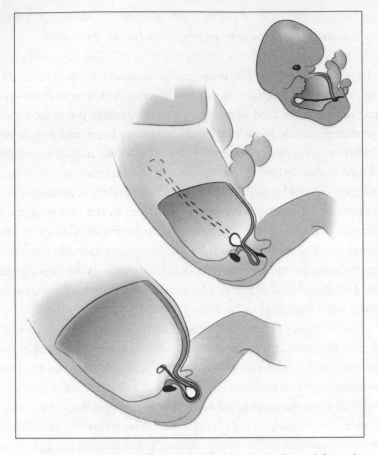

The descent of the testes. During growth, the testes descend from the gonads' primitive position high up in the body. They end up lying in the scrotum, which is an outpocket of the body wall. All of this leaves the body wall of human males weak in the groin area.

through it, their abdominal wall is much stronger than a man's. This is a good thing when you think of the enormous stresses that female body walls go through during pregnancy and childbirth. A tube through the body wall just wouldn't do. Men's tendency to develop hernias is a trade-off between our fish ancestry and our mammal present.

MICROBIAL PAST: MITOCHONDRIAL DISEASES

Mitochondria exist inside every cell of our bodies, doing a remarkable number of things. Their most obvious job is to turn oxygen and sugars into a kind of energy we can use inside our cells. Other functions include metabolizing toxins in our livers and regulating different parts of the function of our cells. We notice our mitochondria only when things go wrong. Unfortunately, the list of diseases caused by malfunctioning mitochondria is extraordinarily long and complex. If there is a problem in the chemical reactions in which oxygen is consumed, energy production can be impaired. The malfunction may be confined to individual tissues, say the eyes, or may affect every system in the body. Depending on the location and severity of the malfunction, it can lead to anything from weakness to death.

Many of the processes we use to live reflect our mitochondria's history. The chain reaction of chemical events that turns sugars and oxygen into usable energy and carbon dioxide arose billions of years ago, and versions of it are still seen in diverse microbes. Mitochondria carry this bacterial past inside of them: with an entire genetic structure and cellular microstructure similar to bacteria, it is generally accepted that they originally arose from free-living microbes over a billion years ago. In fact, the entire energy-generating machinery of our mitochondria arose in one of these kinds of ancient bacteria.

The bacterial past can be used to our advantage in studying the diseases of mitochondria—in fact, some of the best experimental models for these diseases *are* bacteria. This is powerful because we can do all kinds of experiments with bacteria that are not possible with human cells. One of the most provocative studies was done by a team of scientists from Italy and Germany. The disease that they studied invariably kills the infants who are born with it.

Called cardioencephalomyopathy, it results from a genetic change that interrupts the normal metabolic function of mitochondria. In studying a patient who had the disease, the European team identified a place in the DNA that had a suspicious change. Knowing something about the history of life, they then turned to the microbe known as *Paracoccus denitrificans*, which is often called a free-living mitochondrion because its genes and chemical pathways are so similar to those of mitochondria. Just how similar was revealed by the European team. They produced the same change in the bacteria's genes that they saw in their human patient. What they found makes total sense, once we know our history. They were able to simulate parts of a human mitochondrial disease in a bacterium, with virtually the same change in metabolism. This is putting a many-billion-year part of our history to work for us.

The example from microbes is not unique. Judging by the Nobel Prizes awarded in medicine and physiology in the past thirteen years, I should have called this book *Your Inner Fly, Your Inner Worm*, or *Your Inner Yeast*. Pioneering research on flies won the 1995 Nobel Prize in medicine for uncovering a set of genes that builds bodies in humans and other animals. Nobels in medicine in 2002 and 2006 went to people who made significant advances in human genetics and health by studying an insignificant-looking little worm (*C. elegans*). Similarly, in 2001, elegant analyses of yeast (including baker's yeast) and sea urchins won the Nobel in medicine for increasing our understanding of some of the basic biology of all cells. These are not esoteric discoveries made on obscure and unimportant creatures. These discoveries on yeast, flies, worms, and, yes, fish tell us about how our own bodies work, the causes of many of the diseases we suffer, and ways we can develop tools to make our lives longer and healthier.

As a parent of two young children, I find myself spending a lot of time lately in zoos, museums, and aquaria. Being a visitor is a strange experience, because I've been involved with these places for decades, working in museum collections and even helping to prepare exhibits on occasion. During family trips, I've come to realize how much my vocation can make me numb to the beauty and sublime complexity of our world and our bodies. I teach and write about millions of years of history and about bizarre ancient worlds, and usually my interest is detached and analytic. Now I'm experiencing science with my children—in the kinds of places where I discovered my love for it in the first place.

One special moment happened recently with my son at the Museum of Science and Industry in Chicago. We've gone there regularly over the past three years because of his love of trains and the fact that there is a huge model railroad smack in the center of the place. I've spent countless hours at that one exhibit tracing model locomotives on their little trek from Chicago to Seattle. After a number of weekly visits to this shrine for the train-obsessed, Nathaniel and I walked to corners of the museum we had failed to visit during our train-watching ventures or occasional forays to the full-size tractors and planes. In the back of the museum, in the Henry Crown Space Center, model planets hang from the ceiling and space suits lie in cases together with other memorabilia of the space program of the 1960s and 1970s. I was

under the presumption that in the back of the museum I would
see the trivia that didn't make it to the major exhibits up front.
One display consisted of a battered space capsule that you could
walk around and look inside. It didn't look significant; it seemed
way too small and jury-rigged to be anything really important.
The placard was strangely formal, and I had to read it several
times before it dawned on me: here was the original Command
Module from *Apollo 8*, the actual vessel that carried James Lovell,
Frank Borman, and William Anders on humanity's first trip to the
moon and back. This was the spacecraft whose path I followed
during Christmas break in third grade, and here I was thirty-eight
years later with my own son, looking at the real thing. Of course
it was battered. I could see the scars of its journey and subsequent return to earth. Nathaniel was completely disinterested, so
I grabbed him and tried to explain what it was. But I couldn't
speak; my voice became so choked with emotion that I could
barely utter a single word. After a few minutes, I regained my
composure and told him the story of man's trip to the moon.

But the story I can't tell him until he is older is why I became
speechless and emotional. The real story is that *Apollo 8* is a sym-
bol for the power of science to explain and make our universe
knowable. People can quibble over the extent to which the space
program was about science or politics, but the central fact remains
as clear today as it was in 1968: *Apollo 8* was a product of the
essential optimism that fuels the best science. It exemplifies how
the unknown should not be a source of suspicion, fear, or retreat
to superstition, but motivation to continue asking questions and
seeking answers.

Just as the space program changed the way we look at the
moon, paleontology and genetics are changing the way we view
ourselves. As we learn more, what once seemed distant and unat-
tainable comes within our comprehension and our grasp. We live
in an age of discovery, when science is revealing the inner work-

ings of creatures as different as jellyfish, worms, and mice. We are now seeing the glimmer of a solution to one of the greatest mysteries of science—the genetic differences that make humans distinct from other living creatures. Couple these powerful new insights with the fact that some of the most important discoveries in paleontology—new fossils and new tools to analyze them—have come to light in the past twenty years, and we are seeing the truths of our history with ever-increasing precision. Looking back through billions of years of change, everything innovative or apparently unique in the history of life is really just old stuff that has been recycled, recombined, repurposed, or otherwise modified for new uses. This is the story of every part of us, from our sense organs to our heads, indeed our entire body plan.

What do billions of years of history mean for our lives today? Answers to fundamental questions we face—about the inner workings of our organs and our place in nature—will come from understanding how our bodies and minds have emerged from parts common to other living creatures. I can imagine few things more beautiful or intellectually profound than finding the basis for our humanity, and remedies for many of the ills we suffer, nestled inside some of the most humble creatures that have ever lived on our planet.

I naively thought that the Arctic, being far from ringing phones, email, and other trappings of modern civilization, would give me a wonderfully quiet venue to reflect on the year since *Your Inner Fish* was originally published. Unfortunately, this place has a way of swallowing plans whole. We've returned to the *Tiktaalik* sites in Southern Ellesmere Island this summer with the hopes of learning more about the creature and its world. Ted, Farish, and I concocted all kinds of plans to work these fossil zones and to search for others in a nearby fiord. To do this, we calculated our food needs down to the last candy bar, estimated fuel consumption for the entire summer, and planned dates for camp moves to allow us to get our work done.

But now, I'm writing this in the midst of a snowstorm on July 19. My tent is buckling under a blanket of wet snow pushed against the nylon walls by a stiff Arctic wind. Snow can really shut down our operation, making it virtually impossible to find fossils. Even worse, we might have to delay our pending camp move to a promising fossil site that has been in our plans since 2004, but was delayed each successive season for number of different reasons. I'm reminded of a phrase my son's kindergarten teacher uses that works equally well when doling out snacks as it does on senior scientists working in the Arctic, "You get what you get and you don't get upset."

Here in the Arctic, our window into the world of 375 million years ago is a hole in the ground, about 30 feet across and 12 feet deep. Over the years we have removed about 30 cubic feet of rock, mostly by hand, in an effort to expose a very special layer, one loaded with fossil bones. It is a funny scene: the Arctic is a vast expanse of barren ground with little obvious life on the surface. Yet, if you were to look at our work in the quarry, you would find six adults working a small hole so cramped that our heads, shoulders, and legs bump against one another.

The valley where we find *Tiktaalik* (left) is a vast place, yet the actual fossil site (arrow and right) is cramped. Photographs by the author.

We sit long hours with a brush in one hand, a small scratch awl in the other, our faces inches from the rock. Our eyes must be kept close to the rock because the distinction between bone and the surrounding sediments is a fine one. Sometimes the only thing that distinguishes a bone from the rock is an odd sheen or a different texture. Every small clump of dirt or mud may be obscuring a potentially important find. One of our best *Tiktaalik* specimens was originally just a little clump of bone peeking out of the rocks. These kinds of things are scarily easy to miss—I hate to think of how many important specimens we might have overlooked because of wet rock, wind, or even the wrong light conditions on the days we worked.

Since we originally described *Tiktaalik* in 2006, we have used this kind of fieldwork to learn more about the creature. Things have been busy in the lab as well. Fred Mullison and Bob Masek, who you might recall are our expert fossil preparators, have been scratching away to expose the underside of the skull of *Tiktaalik* to reveal parts of the palate and the braincase. Remember Jason Downs, the young college student who joined us in 2000 only to find the main *Tiktaalik* site? Jason is now Dr. Jason P. Downs, Ph.D., and is doing postdoctoral work on these very parts of *Tiktaalik*.

Here, in the underside of the skull, comes our biggest breakthrough since the book was published. As announced in *Nature* magazine in 2008, when Fred and Bob uncovered these bits, we discovered clues to the ways *Tiktaalik* moved, breathed, and supported itself. For the first time we see *Tiktaalik* as an aquatic animal specializing to breathe air and supporting itself on solid ground. These insights come from several important observations, most notably that, unlike every other bony fish, *Tiktaalik* is missing an important plate of bone—the operculum.

The operculum is a plate of bone that forms a flap that covers the gills in most bony fish. You have probably seen the operculum at work in fish that have been trapped out of the water. As they try to breathe this flap opens and closes. The normal operation of the operculum assists the fish in moving water across the gills. Most of them accomplish this feat by using a form of push-pull. Water enters the mouth and throat then, as the throat closes, the operculum opens: water is pushed out of the throat, and pulled across the gills by the opening of the operculum. This fishy way of breathing differs from that seen in land living animals: water or air is moved to the respiratory organs (either gills or lungs) only by mouth pumping or by changes in the shape of the chest cavity. A frog, for example, moves air in and out of its lungs solely by mouth pumping. So, *Tiktaalik* is a finned fish with true gills and lungs, but it has lost the operculum. It relies mostly on mouth breathing, just like land-living animals.

Believe it or not, the loss of the operculum also heralded a change in how *Tiktaalik* moved about. The operculum is one of a series of bones that serve to attach the head to the body in fish. This connection means that when a fish wants to move its head, it needs to move its body. *Tiktaalik* is different. By losing the operculum and all of the bones that would have served to connect the head to the shoulder, *Tiktaalik* has a true neck. This means that it could move its head independently of the body, much like all animals that evolved to the ability to walk on land. Fish swim and feed in three-dimensional space and are readily able to orient the body to position their mouth toward prey. A neck is advantageous in settings where the animal is supporting itself on solid ground, as is the case in shallow pools of water, or on land. To get a sense of how important this is, imagine trying to look around while doing a pushup—you'd be out of luck without a neck.

The underside of the skull also shows just how intermediate *Tiktaalik* is in other ways. The stapes, a bone in our middle ear that we use to hear was originally a bone in the gill arch series in fish. We know this from comparative anatomy and genetics. To make this evolutionary shift, this bone had to substantially reduce in size, from its primitive condition as a large bone supporting the skull in sharks and bony fish, to a tiny bone in the ear. We first see a stapes in early amphibians such as *Acanthostega*. Look to related lobe-finned fish, and we still see a large bone, shaped almost like a boomerang. One of the best known of these fish, *Eusthenopteron*, has a large bone with a number of joints that suggest it was a major link between different bones of the skull. We now know what this bone looks like in *Tiktaalik*. Fred and Bob removed the bones from the fish and guess what? *Tiktaalik* has a bone smaller than *Eusthenopteron*, yet larger than the same bone in *Acanthostega*. It is beautifully intermediate in size.

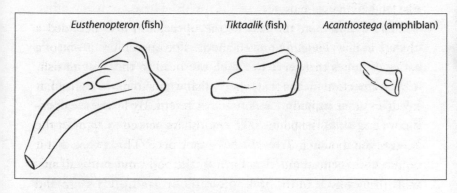

Eusthenopteron (fish)   Tiktaalik (fish)   Acanthostega (amphibian)

A gill arch bone, the hyomandibula, gradually becomes smaller over millions of years to be a tiny bone in our middle ear. This bone in *Tiktaalik* is intermediate in size between related lobe-finned fish and early amphibians. Art by Kalliopi Monoyios.

But why stop at ears? The head of lobe-finned fish such as *Eusthenopteron* has an odd joint in the middle of the braincase. Oddly, the front end of the skull can flex relative to the back end. The skull of land-living animals, such as *Acanthostega*, is more rigid: the joint is lost and the bones are firmly sutured together. What does *Tiktaalik* have? A joint is present in the same place as in *Eusthenopteron*. But this joint has extremely restricted motions, greatly reduced from the primitive condition seen in *Eusthenopteron*. Other parts of the braincase are very fishlike, such as the shape of the rear end, or very like amphibians, such as the shape of the palate.

Since *Your Inner Fish* was published in January 2008, I've been asked a number of questions about our distant past and the ways we paleontologists reconstruct it. By far the most common question is about global warming, "Do a warming Arctic and retreating glaciers give us more rock to look at?" The answer from my perspective is no. The rocks we work on are exposed to the same extent today as they were when Ashton Embry first walked them as part of his mapping project in the 1970s. There are many other changes, though. Certainly the biggest of these is the sheer amount of work that is going on up in the Arctic nowadays. With rising oil, gas, and mineral prices, Arctic exploration is occurring at a much more frenzied pace than when we first started. Our current campsite gives us an important lesson in this regard. We've returned to a site we originally visited in 2000. Seeing the place gave us a shocking revelation—our own footprints were still recorded in the Arctic tundra up to eight years later. I could even make out my own boot print from a walk I took back to camp one day in July 2000. We must tread carefully in this very special and fragile ecosystem.

People also ask how *Tiktaalik* could have survived life in the Arctic climate. Look out from the *Tiktaalik* quarry (known more officially as NV2K17) and you see a classic polar landscape. Rivers drain massive glaciers four miles upstream, Arctic foxes, wolves, and musk oxen dwell in the valley, and snow patches are present even in midsummer. It is cold. Yet the world my colleagues and I are exposing is a tropical one—with warm water fish and plants. Tropical fossils inside Arctic rocks can mean two things: global climate change and/or moving continents. In this case we have both: the climate today is different from that 375 million years ago and the rocks that are in Ellesmere Island today were once very close to the equator.

Another question I often get is whether *Tiktaalik* is the "missing link." Paleontologists have real trouble with this term. Not the least of these problems is that *Tiktaalik* is a *found* link. But the difficulty goes deeper. It helps to consider what links we have between lobe finned fish and land living animals. We have the DNA that clearly shows how lobe finned fish are genetically closely related amphibians. We also now know about the workings of some of that DNA, the DNA that builds bodies. It turns out that much of the genetic recipe that builds fins is similar to that of limbs. Then, there are the features we can compare among living fish and amphibians. Lobe-finned fish, such as lungfish are so similar to amphibians that they have often been confused with salamanders. And finally there are the fossils, whole batches of them with names such as those we discussed in the book like *Eusthenopteron*, *Panderichthys*, *Acanthostega*, and *Ichthyostega*, and those described since its publication, such as *Gogonasus* and *Ventastega*, to name a few. These are all close relatives of one another, more like cousins to different degrees rather than direct ancestors and descendents. The whole series shows intermediate conditions in head shape, limb bones, shoulders, hips, and other structures. Given all of this, *Tiktaalik* is not the missing link,

July 19, 2008, in Bird Fiord. Photograph by author.

it is one in a series of found links. And there are more to discover. This is the reason for my return to the Arctic. And the reason why I'll continue to return there.

Crazy as it sounds, I'm going to be quite sad when, in a few days, I say goodbye to this valley and this hole in the ground. Ted, Farish, Jason, Fred, and I have worked almost a decade to find and study bones from this little spot. We've dug through layer after layer of geological strata as part of our search. There are layers of personal history in these rocks too, with all the travails, joys, and lessons learned here over the years. But we are moving on. On to younger rocks and perhaps, if we are really lucky, the discovery of "*Tiktaalik* 2.0," then "*Tiktaalik* 3.0," and so on. With each new fossil find, we discover answers to old questions and are challenged to pose new, more refined, ones. That is the thrill of the hunt.

July 2008
Bird Fiord,
Southern Ellesmere Island

CHAPTER ONE   FINDING YOUR INNER FISH

I have included a mix of primary and secondary sources for those interested in pursuing the topics in the book further. For accounts that use exploratory paleontological expeditions as a vehicle to discuss major questions in biology and geology, see Mike Novacek's *Dinosaurs of the Flaming Cliffs* (New York: Anchor, 1997), Andrew Knoll's *Life on a Young Planet* (Princeton: Princeton University Press, 2002), and John Long's *Swimming in Stone* (Melbourne: Freemantle Press, 2006). All balance scientific analysis with descriptions of discovery in the field.

The comparative methods that I discuss, including the methods used in our walk through the zoo, are the methods of cladistics. A superb overview is Henry Gee's *In Search of Deep Time* (New York: Free Press, 1999). Basically, I present a version of the three-taxon statement, the starting point for cladistic comparisons. A good treatment with background sources is found in Richard Forey et al., "The Lungfish, the Coelacanth and the Cow Revisited," in H.-P. Schultze and L. Trueb, eds., *Origin of the Higher Groups of Tetrapods* (Ithaca, N.Y.: Cornell University Press, 1991).

The correlation between the fossil record and our "walk through the zoo" is discussed in several papers. A sampling: Benton, M. J., and Hitchin, R. (1997) Congruence between phylogenetic and stratigraphic data in the history of life, *Proceedings of the Royal Society of London,* B 264:885–890; Norell, M. A., and Novacek, M. J. (1992) Congruence between superpositional and phylogenetic patterns: Comparing cladistic patterns with fossil records, *Cladistics* 8:319–337; Wagner, P. J., and Sidor, C. (2000) Age rank/clade rank metrics—sampling, taxonomy, and the meaning of "stratigraphic consistency," *Systematic Biology* 49:463–479.

The layers of the rock column and the fossils contained therein are beautifully and comprehensively discussed in Richard Fortey's *Life: A Natural History of the*

First Four Billion Years of Life on Earth (New York: Knopf, 1998). Resources for vertebrate paleontology include R. Carroll, *Vertebrate Paleontology and Evolution* (San Francisco: W. H. Freeman, 1987), and M. J. Benton, *Vertebrate Paleontology* (London: Blackwell, 2004).

For the origin of tetrapods: Carl Zimmer reviewed the state of the art in the field in his highly readable and well-researched *At the Water's Edge* (New York: Free Press, 1998). Jenny Clack has written the definitive text on the whole transition, *Gaining Ground* (Bloomington: Indiana University Press, 2002). The bible of this transition, Clack's book will bring a novice to expert status quickly.

Our original papers describing *Tiktaalik* appeared in the April 6, 2006, issue of *Nature.* The references are: Daeschler et al. (2006) A Devonian tetrapod-like fish and the origin of the tetrapod body plan, *Nature* 757:757–763; Shubin et al. (2006) The pectoral fin of *Tiktaalik roseae* and the origin of the tetrapod limb, *Nature* 757:764–771. Jenny Clack and Per Ahlberg had a very readable and comprehensive commentary piece in the same issue (*Nature* 757:747–749).

Everything about our past is relative, even the structure of this book. I could have called this book "Our Inner Human" and written it from a fish's point of view. The structure of that book would have been strangely similar: a focus on the history humans and fish share in bodies, brains, and cells. As we've seen, all life shares a deep part of its history with other species, while another part of its history is unique.

CHAPTER TWO   GETTING A GRIP

Owen was by no means the first person to see the pattern of one bone–two bones–lotsa blobs–digits. Vicq-d'Azyr in the 1600s and Geoffroy St. Hilaire (1812) also made this pattern part of their worldviews. What distinguished Owen was his concept of the archetype. This was a transcendental organization of the body, reflecting the design of the Creator. St. Hilaire was searching less for an archetypical pattern hidden in all structure than for "laws of form" that govern the formation of bodies. A nice treatment of these issues is in T. Appel, *The Cuvier-Geoffroy Debate: French Biology in the Decades Before Darwin* (New York: Oxford University Press, 1987), and E. S. Russell, *Form and Function: A Contribution to the History of Morphology* (Chicago: University of Chicago Press, 1982).

A recent volume edited by Brian Hall is one-stop shopping for information on limb diversity and development and contains a number of important papers on different kinds of limbs: Brian K. Hall, ed., *Fins into Limbs: Evolution, Development, and Transformation* (Chicago: University of Chicago Press, 2007). Useful references for exploring the shift from fins and limbs in more detail include Shubin et al. (2006) The pectoral fin of *Tiktaalik roseae* and the origin of the tetrapod limb, *Nature* 757:764–771; Coates, M. I., Jeffery, J. E., and Ruta, M. (2002) Fins to limbs: what the fossils say, *Evolution and Development* 4:390–412.

CHAPTER THREE   HANDY GENES

The developmental biology of limb diversity has seen a number of reviews and primary papers. For a review of the classic literature see Shubin, N., and Alberch, P. (1986) A morphogenetic approach to the origin and basic organization of the tetrapod limb, *Evolutionary Biology* 20:319–387; and Hinchliffe, J. R., and Griffiths, P., "The Pre-chondrogenic Patterns in Tetrapod Limb Develoment and Their Phylogenetic Significance," in B. Goodwin, N. Holder, and C. Wylie, eds., *Development and Evolution* (Cambridge, Eng.: Cambridge University Press, 1983), pp. 99–121. Saunders's and Zwilling's experiments are now classic, so some of the best accounts are now seen in the major textbooks in developmental biology. These include S. Gilbert, *Developmental Biology*, 8th ed. (Sunderland, Mass.: Sinauer Associates, 2006); L. Wolpert, J. Smith, T. Jessell, F. Lawrence, E. Robertson, and E. Meyerowitz, *Principles of Development* (Oxford, Eng.: Oxford University Press, 2006).

A note on what I mean when I use the term "primitive." We tend to use this term to describe features of organisms, not the organisms themselves. That is, fins are primitive to limbs, but fish are not primitive to people. The reason for this is that living fish are highly specialized animals in their own right, just like we are. I occasionally call creatures primitive as a shorthand for the notion that they are similar to the ancestral condition in whatever feature I'm describing.

For the first paper describing *Sonic hedgehog*'s role in limb patterning, go to Riddle, R., Johnson, R. L., Laufer, E., Tabin, C. (1993) *Sonic hedgehog* mediates the polarizing activity of the ZPA, *Cell* 75:1401–1416.

Randy's work on *Sonic* signaling in shark and skate fins is in Dahn, R., Davis, M., Pappano, W., Shubin, N. (2007) *Sonic hedgehog* function in chondrichthyan fins and the evolution of appendage patterning, *Nature* 445:311–314. Subsequent work from the lab on the origin of limbs, at least from a genetic perspective, is contained in Davis, M., Dahn, R., and Shubin, N. (2007) A limb autopodial-like pattern of *Hox* expression in a basal actinopterygian fish, *Nature* 447:473–476.

The stunning genetic similarities in the development of flies, chickens, and humans is discussed in Shubin, N., Tabin, C., Carroll, S. (1997) Fossils, genes, and the evolution of animal limbs, *Nature* 388:639–648; and Erwin, D. and Davidson, E. H. (2003) The last common bilaterian ancestor, *Development* 129: 3021–3032.

CHAPTER FOUR   TEETH EVERYWHERE

The importance of teeth to an understanding of mammals is evident in the many treatments in the field. Dental structure plays a particularly important role in understanding the early record of mammals. Extensive reviews are found in Z. Kielan-Jaworowska, R. L. Cifelli, and Z. Luo, *Mammals from the Age of Dinosaurs* (New York: Columbia University Press, 2004); and J. A. Lillegraven, Z. Kielan-Jaworowska, and W. Clemens, eds., *Mesozoic Mammals: The First Two-Thirds of Mammalian History* (Berkeley: University of California Press, 1979), p. 311.

Farish's mammal from Arizona is analyzed in Jenkins, F. A., Jr., Crompton, A. W., Downs, W. R. (1983) Mesozoic mammals from Arizona: New evidence on mammalian evolution, *Science* 222:1233–1235.

The tritheledonts we found in Nova Scotia are described in Shubin, N., Crompton, A. W., Sues, H.-D., and Olsen, P. (1991) New fossil evidence on the sister-group of mammals and early Mesozoic faunal distributions, *Science* 251:1063–1065.

A recent review on the origin of teeth, bone, and skulls, in particular the new evolution gleaned from conodont animals, is found in Donoghue, P., and Sansom I. (2002) Origin and early evolution of vertebrate skeletonization, *Microscopy Research and Technique* 59:352–372. A thorough review of the evolutionary relationships among conodonts and their significance is in Donoghue, P., Forey, P., and Aldridge, R. (2000) Conodont affinity and chordate phylogeny, *Biological Reviews* 75:191–251.

CHAPTER FIVE   GETTING AHEAD

A wonderfully comprehensive and detailed treatment of the details of skull structure, development, and evolution is found in a three-volume set: *The Skull*, James Hanken and Brian Hall, eds. (Chicago: University of Chicago Press, 1993). This is a multi-author update of one of the classic volumes on head development and structure: G. R. de Beer, *The Development of the Vertebrate Skull* (Oxford, Eng.: Oxford University Press, 1937).

Details of head development and structure in humans can be found in texts on human anatomy and embryology. For embryology, see K. Moore and T.V.N. Persaud, *The Developing Human*, 7th ed. (Philadelphia: Elsevier, 2006). The companion anatomy text is K. Moore and A. F. Dalley, *Clinically Oriented Anatomy* (Philadelphia: Lippincott Williams & Wilkins, 2006).

Francis Maitland Balfour's seminal work is encapsulated in Balfour, F. M.

(1874) A preliminary account of the development of the elasmobranch fishes, *Q. J. Microsc. Sci.* 14:323–364; F. M. Balfour, *A Monograph on the Development of Elasmobranch Fishes*, 4 vols. (London: Macmillan & Co., 1878); F. M. Balfour, *A Treatise on Comparative Embryology*, 2 vols. (London: Macmillan & Co., 1880–81); M. Foster and A. Sedgwick, *The Works of Francis Maitland Balfour*, with an introductory biographical notice by Michael Foster, 4 vols. (London: Macmillan & Co., 1885). A successor at Oxford, Edwin Goodrich, produced one of the classics of comparative anatomy, *Studies on the Structure and Development of Vertebrates* (London: Macmillan, 1930).

Balfour, Oken, Goethe, Huxley, and others were addressing the problem known as head segmentation. Just as the vertebrae differ in a regular progression from front to back, so the head has a segmental pattern. A selection of classic and recent resources (all with good bibliographies) to pursue this field further: Olsson, L., Ericsson, R., Cerny, R. (2005) Vertebrate head development: Segmentation, novelties, and homology, *Theory in Biosciences* 124:145–163; Jollie, M. (1977) Segmentation of the vertebrate head, *American Zoologist* 17:323–333; Graham, A. (2001) The development and evolution of the pharyngeal arches, *Journal of Anatomy* 199:133–141.

A recent overview of the genetic basis of gill arch formation is found in Kuratani, S. (2004) Evolution of the vertebrate jaw: comparative embryology and molecular developmental biology reveal the factors behind evolutionary novelty, *Journal of Anatomy* 205:335–347. Examples of the experimental manipulation of one gill arch into another, using genetic technologies, include Baltzinger, M., Ori, M., Pasqualetti, M., Nardi, I., Riji, F. (2005) *Hoxa2* knockdown in *Xenopus* results in hyoid to mandibular homeosis, *Developmental Dynamics* 234:858–867; Depew, M., Lufkin, T., Rubenstein, J. (2002) Specification of jaw subdivisions by *Dlx* genes, *Science* 298:381–385.

A comprehensive, well-illustrated, and informative resource for early fossil records of skulls, heads, and primitive fish is reviewed in P. Janvier, *Early Vertebrates* (Oxford, Eng.: Oxford University Press, 1996). The paper describing *Haikouella*, the 530-million-year-old worm with gills, is Chen, J.-Y., Huang, D. Y., and Li, C. W. (1999) An early Cambrian craniate-like chordate, *Nature* 402:518–522.

CHAPTER SIX   THE BEST-LAID (BODY) PLANS

The origin of body plans has been the subject of a number of book-length treatments. For one with an exceptional scope and bibliography, go to J. Valentine, *On the Origin of Phyla* (Chicago: University of Chicago Press, 2004).

There have been several biographies of von Baer. A short one is Jane Oppenheimer, "Baer, Karl Ernst von," in C. Gillespie, ed., *Dictionary of Scientific Biog-*

raphy, vol. 1 (New York: Scribners, 1970). For more detailed treatments, see *Autobiography of Dr. Karl Ernst von Baer*, ed. Jane Oppenheimer (1986; originally published in German, 2nd ed., 1886). See also B. E. Raikov, *Karl Ernst von Baer, 1792–1876*, trans. from Russian (1968), and Ludwig Stieda, *Karl Ernst von Baer*, 2nd ed. (1886). All these resources have large bibliographies. See also S. Gould, *Ontogeny and Phylogeny* (Cambridge, Mass.: Harvard University Press, 1977), for a discussion of von Baer's laws.

Spemann and Mangold's experiments are discussed in embryology textbooks: S. Gilbert, *Developmental Biology*, 8th ed. (Sunderland, Mass.: Sinauer Associates, 2006). A modern genetic perspective on the Organizer is contained in De Robertis, E. M. (2006) Spemann's organizer and self regulation in amphibian embryos, *Nature Reviews* 7:296–302, and De Robertis, E. M., and Arecheaga, J. The Spemann Organizer: 75 years on, *International Journal of Developmental Biology* 45 (special issue).

For access to the huge literature on *Hox* genes and evolution, the best starting reference is Sean Carroll's recent book *Endless Forms Most Beautiful* (New York: Norton, 2004). A recent review and interpretation of the ways that genes allow us to understand the common ancestor of bilaterally symmetrical animals is in Erwin, D., and Davidson, E. H. (2002) The last common bilaterian ancestor, *Development* 129:3021–3032.

A number of investigators argue that a genetic "flip" between the body plan of an anthropod and the body plan of a human happened sometime in the distant past. This idea is discussed in De Robertis, E., and Sasai, Y. (1996) A common plan for dorsoventral patterning in Bilateria, *Nature* 380:37–40. Historical perspective on St. Hilaire's views, as well as other controversies in the early years of comparative anatomy, are found in T. Appel, *The Cuvier-Geoffroy Debate: French Biology in the Decades Before Darwin* (New York: Oxford University Press, 1987). Data from acorn worms does not easily fit this model, and suggests that in some taxa the map between gene activity and axis specification may have evolved. For this work, see Lowe, C. J., et al. (2006) Dorsoventral patterning in hemichordates: insights into early chordate evolution, *PLoS Biology online access:* http://dx.doi.org/journal .0040291.

The evolution of the genes that determine the body axes is reviewed in Martindale, M. Q. (2005) The evolution of metazoan axial properties, *Nature Reviews Genetics* 6:917–927. Body plan genes in cnidarians (jellyfish, sea anemones, and their relatives) are discussed in a series of primary papers: Martindale, M. Q., Finnerty, J. R., Henry, J. (2002) The Radiata and the evolutionary origins of the bilaterian body plan, *Molecular Phylogenetics and Evolution* 24:358–365; Matus, D. Q., Pang, K., Marlow, H., Dunn, C., Thomsen, G., Martindale, M. (2006) Molecular evidence for deep evolutionary roots of bilaterality in animal development, *Proceedings of the National Academy of Sciences* 103:11195–11200; Chourrout, D., et al. (2006) Minimal protohox cluster inferred from bilaterian and cnidarian *Hox* complements, *Nature* 442:684–687; Martindale, M., Pang, K., Finnerty, J. (2004) Investigating the origins of triploblasty: "mesodermal" gene expression in a diploblastic animal, the sea anemone *Nemostella vectensis* (phylum, Cnidaria; class, Anthozoa), *Development* 131:2463–2474; Finnerty, J., Pang, K., Burton, P., Paulson, D., Martindale, M. Q. (2004) Deep origins for bilateral symmetry: *Hox* and *Dpp* expression in a sea anemone, *Science* 304:1335–1337.

CHAPTER SEVEN ADVENTURES IN BODYBUILDING

Three key articles review the origins and evolution of bodies and offer an integrative perspective on genetics, geology, and ecology: King, N. (2004) The unicellular ancestry of animal development, *Developmental Cell* 7:313–325; Knoll, A. H., and Carroll, S. B. (1999) Early animal evolution: Emerging views from comparative biology and geology, *Science* 284:2129–2137; Brooke, N. M., and Holland, P. (2003) The evolution of multicellularity and early animal genomes, *Current Opinion in Genetics and Development* 13:599–603. All three papers are well referenced and offer a good introduction to the topics of the chapter.

For stimulating treatments of the consequences of the origin of bodies and of other new forms of biological organization, see L. W. Buss, *The Evolution of Individuality* (Princeton: Princeton University Press, 2006), and J. Maynard Smith, and E. Szathmary, *The Major Transitions in Evolution* (New York: Oxford University Press, 1998).

The story behind the Ediacarian animals is covered, with references, in Richard Fortey's *Life: A Natural History of the First Four Billion Years of Life on Earth* (New York: Knopf, 1998), and Andrew Knoll's *Life on a Young Planet* (Princeton: Princeton University Press, 2002).

The experiment that yielded "proto-bodies" from "no-bodies" is described in Boraas, M. E., Seale, D. B., Boxhorn, J. (1998) Phagotrophy by a flagellate selects for colonial prey: A possible origin of multicellularity, *Evolutionary Ecology* 12:153–164.

CHAPTER EIGHT MAKING SCENTS

The University of Utah has an effective website, Learn.Genetics, that provides a wonderfully simple kitchen protocol for extracting DNA. The URL is http://learn.genetics.utah.edu/units/activities/extraction/.

The evolution of the so-called odor genes or, more precisely, olfactory receptor genes has a large literature. Buck and Axel's seminal paper is Buck, L., and Axel, R.

(1991) A novel multigene family may encode odorant receptors: a molecular basis for odor recognition, *Cell* 65:175–181.

Comparative aspects of olfactory gene evolution are treated in Young, B., and Trask, B. J. (2002) The sense of smell: genomics of vertebrate odorant receptors, *Human Molecular Genetics* 11:1153–1160; Mombaerts, P. (1999) Molecular biology of odorant receptors in vertebrates, *Annual Reviews of Neuroscience* 22:487–509.

Olfactory receptor genes in jawless fish are discussed in Freitag, J., Beck, A., Ludwig, G., von Buchholtz, L., Breer, H. (1999) On the origin of the olfactory receptor family: receptor genes of the jawless fish (*Lampetra fluviatilis*), *Gene* 226:165–174. The distinction between aquatic and terrestrial olfactory receptor genes is described in Freitag, J., Ludwig, G., Andreini, I., Rossler, P., Breer, H. (1998) Olfactory receptors in aquatic and terrestrial vertebrates, *Journal of Comparative Physiology A* 183:635–650.

Human olfactory receptor evolution is discussed in a number of papers. This selection reflects the issues discussed in the text: Gilad, Y., Man, O., Lancet, D. (2003) Human specific loss of olfactory receptor genes, *Proceedings of the National Academy of Sciences* 100:3324–3327; Gilad, Y., Man, O., and Glusman, G. (2005) A comparison of the human and chimpanzee olfactory receptor gene repertoires, *Genome Research* 15:224–230; Menashe, I., Man, O., Lancet, D., Gilad, Y. (2003) Different noses for different people, *Nature Genetics* 34:143–144; Gilad, Y., Wiebe, V., Przeworski, M., Lancet, D., Paabo, S. (2003) Loss of olfactory receptor genes coincides with the acquisition of full trichromatic vision in primates, *PLoS Biology online access:* http://dx.doi.org/journal.pbio.0020005.

The notion of gene duplication as an important source of new genetic variation traces to the seminal work of Ohno almost forty years ago: S. Ohno, *Evolution by Gene Duplication* (New York: Springer-Verlag, 1970). A recent review of the issue that contains a discussion of both opsins and olfactory receptor genes is found in Taylor, J., and Raes, J. (2004) Duplication and divergence: the evolution of new genes and old ideas, *Annual Reviews of Genetics* 38:615–643.

CHAPTER NINE   VISION

Opsin genes in the evolution of eyes have been described in a number of papers in recent years. Reviews of the basic biology and the consequences of opsin gene evolution include Nathans, J. (1999) The evolution and physiology of human color vision: insights from molecular genetic studies of visual pigments, *Neuron* 24:299–312; Dominy, N., Svenning, J. C., Li, W. H. (2003) Historical contingency in the evolution of primate color vision, *Journal of Human Evolution* 44:25–45; Tan, Y., Yoder, A., Yamashita, N., Li, W. H. (2005) Evidence from opsin genes rejects nocturnality in ancestral primates, *Proceedings of the National Acad-*

emy of Sciences 102:14712–14716; Yokoyama, S. (1996) Molecular evolution of retinal and nonretinal opsins, *Genes to Cells* 1:787–794; Dulai, K., von Dornum, M., Mollon, J., Hunt, D. M. (1999) The evolution of trichromatic color vision by opsin gene duplication in New World and Old World primates, *Genome* 9:629–638.

Detlev Arendt and Joachim Wittbrodt's work on photoreceptor tissues was originally described in a paper from the primary literature: Arendt, D., Tessmar-Raible, K., Synman, H., Dorresteijn, A., Wittbrodt, J. (2004) Ciliary photoreceptors with a vertebrate-type opsin in an invertebrate brain, *Science* 306:869–871. An associated commentary appeared with the piece: Pennisi, E. (2004) Worm's light-sensing proteins suggest eye's single origin, *Science* 306:796–797. An earlier review by Arendt provides the larger framework that he uses to interpret the discovery: Arendt, D. (2003) The evolution of eyes and photoreceptor cell types, *International Journal of Developmental Biology* 47:563–571. Further commentary can be found in Plachetzki, D. C., Serb, J. M., Oakley, T. H. (2005) New insights into photoreceptor evolution, *Trends in Ecology and Evolution* 20:465–467. Still more commentary on Arendt and Wittbrodt's work by Bernd Fritzsch and Joram Piatigorsky appeared in a later issue of *Science,* with a comment-reply that discussed the notion that the origin of eyes may be extremely ancient, and traced to a very deep branch of our evolutionary tree. This text can be found in *Science* (2005) 308:1113–1114.

A review of Walter Gehring's work on *Pax 6* and its consequences for eye evolution is contained in a personal account: Gehring, W. (2005) New perspectives on eye development and the evolution of eyes and photoreceptors, *Journal of Heredity* 96:171–184.

Papers that look at the different possible relationships between conserved eye formation genes and the evolution of eye organs include Oakley, T. (2003) The eye as a replicating and diverging modular developmental unit, *Trends in Ecology and Evolution* 18:623–627, and Nilsson D.-E. (2004) Eye evolution: a question of genetic promiscuity, *Current Opinion in Neurobiology* 14:407–414.

The relationship between the lens proteins in our eyes and those of larval sea squirts is discussed in Shimeld, S., Purkiss, A. G., Dirks, R.P.H., Bateman, O., Slingsby, C., Lubsen, N. (2005) Urochordate by-crystallin and the evolutionary origin of the vertebrate eye lens, *Current Biology* 15:1684–1689.

CHAPTER TEN   EARS

The genetics of inner ear evolution is discussed in Beisel, K. W., and Fritzsch, B. (2004) Keeping sensory cells and evolving neurons to connect them to the brain: molecular conservation and novelties in vertebrate ear development,

Brain Behavior and Evolution 64:182–197. Ear development and the genes behind it are discussed in Represa, J., Frenz, D. A., Van de Water, T. (2000) Genetic patterning of embryonic ear development, *Acta Otolaryngolica* 120:5–10.

The transformation of the hyomandibula into the stapes is reviewed in comprehensive book-length treatments of the evolution of primitive fish or the origin of land-living animals: J. Clack, *Gaining Ground* (Bloomington: Indiana University Press, 2002); P. Janvier, *Early Vertebrates* (Oxford, Eng.: Oxford University Press, 1996). It is also discussed in recent research papers, including Clack, J. A. (1989) Discovery of the earliest known tetrapod stapes, *Nature* 342:425–427; Brazeau, M., and Ahlberg, P. (2005) Tetrapod-like middle ear architecture in a Devonian fish, *Nature* 439:318–321.

The origin of the mammalian middle ear is discussed from the perspective of a scientific historian in P. Bowler, *Life's Spendid Journey* (Chicago: University of Chicago Press, 1996). Key primary sources include: Reichert, C. (1837) Uber die Visceralbogen der Wirbeltiere im allgemeinen und deren Metamorphosen bei den Vogeln und Saugetieren, *Arch. Anat. Physiol. Wiss. Med.* 1837:120–222; Gaupp, E. (1911) Beiträge zur Kenntnis des Unterkiefers der Wirbeltiere I. Der Processus anterior (Folii) des Hammers der Sauger und das Goniale der Nichtsäuger, *Anatomischer Anzeiger* 39:97–135; Gaupp, E. (1911) Beiträge zur Kenntnis des Unterkiefers der Wirbeltiere II. Die Zusammensetzung des Unterkiefers der Quadrupeden, *Anatomischer Anzeiger,* 39:433–473; Gaupp, E. (1911) Beiträge zur Kenntnis des Unterkiefers der Wirbeltiere III. Das Probleme der Entstehung eines "sekundären" Kiefergelenkes bei den Säugern, *Anatomischer Anzeiger,* 39:609–666; Gregory, W. K. (1913) Critique of recent work on the morphology of the vertebrate skull, especially in relation to the origin of mammals, *Journal of Morphology* 24:1–42.

Major literature on the origin of the mammalian jaw, chewing, and the three-boned middle ear includes Crompton, A. W. (1963) The evolution of the mammalian jaw, *Evolution* 17:431–439; Crompton, A. W., and Parker, P. (1978) Evolution of the mammalian masticatory apparatus, *American Scientist* 66:192–201; Hopson, J. (1966) The origin of the mammalian middle ear, *American Zoologist* 6:437–450; Allin, E. (1975) Evolution of the mammalian ear, *Journal of Morphology* 147:403–438.

The evolutionary origin of *Pax 2* and *Pax 6* and the evolutionary link of ears and eyes to box jellyfish is discussed in Piatigorsky, J., and Kozmik, Z. (2004) Cubozoan jellyfish: an evo/devo model for eyes and other sensory systems, *International Journal for Developmental Biology* 48:719–729.

Links of sensory receptor molecules to different molecules in bacteria are discussed in Kung, C. (2005) A possible unifying principle for mechanosensation, *Nature* 436:647–654.

CHAPTER ELEVEN   THE MEANING OF IT ALL

Biologists now use the family tree of life as the basis for classification. This makes sense, as the names for our groups of animals have biological meaning—they reflect their ancestry. If we use the tree of life in this way, groups, like phyla, families, genera, etc., would include everything on one branch. To use the family analogy again, this is akin to saying that everybody who has descended from my dad's parents are Shubins. Nobody is arbitrarily excluded. The challenge for non-specialists comes about when we use the family tree of life as the basis for our taxonomy, many of the names we are familiar with cease to have scientific meaning. Take, reptiles, for example. We all learn what reptiles are from our trips to the zoo, they include things like lizards, snakes, turtles, dinosaurs, and crocodiles, to name a few. But it turns out that that the branch of the tree of life that includes all of these creatures also includes other animals, such as mammals and birds. Reptiles include some, but not all members of a branch of the tree of life, something that is just as arbitrary as excluding a blood relative from one's family.

The methods of phylogenetic systematics are discussed in a number of sources. Key primary literature includes the classic work of Willi Hennig, published originally in German (*Grundzüge einer Theorie der phylogenetischen Systematik* [Berlin: Deutscher Zentralverlag, 1950]) and translated into English more than a decade later (*Phylogenetic Systematics*, trans. D. D. Davis and R. Zangerl [Urbana: University of Illinois Press, 1966]).

Methods of phylogenetic reconstruction, which form the basis for the chapter, are discussed in detail in P. Forey, ed., *Cladistics: A Practical Course in Systematics* (Oxford, Eng.: Clarendon Press, 1992); D. Hillis, C. Moritz, and B. Mable, eds., *Molecular Systematics* (Sunderland, Mass.: Sinauer Associates, 1996); R. DeSalle, G. Girbet, and W. Wheeler, *Molecular Systematics and Evolution: Theory and Practice* (Basel: Birkhauser Verlag, 2002).

A comprehensive treatment of the phenomenon of independent evolution of similar features is in M. Sanderson and L. Hufford, *Homoplasy: The Recurrence of Similarity in Evolution* (San Diego: Academic Press, 1996).

To see the tree of life and the different hypotheses for the relationships between living creatures, visit http://tolweb.org/tree/.

The notion that our evolutionary history has medical implications has been the subject of several good recent books. For comprehensive and well-referenced treatments, see N. Boaz, *Evolving Health: The Origins of Illness and How the Modern World Is Making Us Sick* (New York: Wiley, 2002); D. Mindell, *The Evolving World: Evolution in Everyday Life* (Cambridge, Mass.: Harvard University Press,

2006); R. M. Nesse and G. C. Williams, *Why We Get Sick: The New Science of Darwinian Medicine* (New York: Vintage, 1996); W. R. Trevathan, E. O. Smith, and J. J. McKenna, *Evolutionary Medicine* (New York: Oxford University Press, 1999).

The apnea example I derived from discussions with Nino Ramirez, chairman of the Department of Anatomy at the University of Chicago. The hiccup example is derived from Straus, C., et al. (2003) A phylogenetic hypothesis for the origin of hiccoughs, *Bioessays* 25:182–188. The human-bacterial gene switch used in the study of mitochondrial cardioencephalomyopathy was originally discussed in Lucioli, S., et al. (2006) Introducing a novel human mtDNA mutation into the *Paracoccus denitriticans* COX 1 gene explains functional deficits in a patient, *Neurogenetics* 7:51–57.

ONLINE RESOURCES

A number of websites and blogs carry accurate information and are updated frequently.

http://www.ucmp.berkeley.edu/ Produced by the Museum of Paleontology at the University of California–Berkeley, this is one of the best online resources on paleontology and evolution. It is continuously updated and revised.

http://www.scienceblogs.com/loom/ This is Carl Zimmer's blog, a well-written, timely, and thoughtful source of information and discussion on evolution.

http://www.scienceblogs.com/pharyngula/ P. Z. Myers, a professor of developmental biology, writes this accessible, informative, and cutting-edge blog. This is a rich source of information, well worth following.

Both Zimmer's and Myers's blogs are at http://www.scienceblogs.com, a site that contains a number of excellent blogs also worth following for information and commentary on recent discoveries. Blogs relevant to the theme of this book at that site include Afarensis, Tetrapod Zoology, Evolving Thoughts, and Gene Expression.

http://www.tolweb.org/tree/ The Tree of Life Project provides a regularly updated and authoritative treatment of the relationships among all groups of life. Like the UCMP page at Berkeley, it also includes resources for learning about how evolutionary trees are made and interpreted.

ACKNOWLEDGMENTS

All the illustrations, except where noted, are by Ms. Kalliopi Monoyios (www.kalliopimonoyios.com). Kapi read drafts of the manuscript and not only improved the text but designed art that matched it. I have been truly fortunate to work with someone with so many talents. Scott Rawlins (Arcadia University) generously gave permission to use his elegant rendering of *Sauripterus* in Chapter 2. Ted Daeschler (Academy of Natural Sciences of Philadelphia) graciously provided his superb photos of the great *Tiktaalik* "C" specimen. Thanks are due to Phillip Donoghue (University of Bristol) and Mark Purnell (University of Leicester) for permission to use their rendering of the conodont tooth array, McGraw-Hill for permission to use the textbook figure that started the hunt for *Tiktaalik,* and Steven Campana of the Canadian Shark Research Laboratory for the photos of shark organs.

One of the greatest debts students of anatomy have is to the people who donate their bodies so that we can learn. It is a rare privilege to learn from a real body. Sitting through long hours in the lab, one feels a very profound connection to the donors who make the experience possible. I felt that connection again while writing this book.

The ideas I present here are rooted in research I've done and in classes I've taught. Colleagues and students too numerous to name—undergraduates, medical students, and graduate students—have played a role in the thinking that went into these pages.

I owe a great debt to the colleagues I have worked with over the years. Ted Daeschler, Farish A. Jenkins, Jr., Fred Mullison, Paul Olsen, William Amaral, Jason Downs, and Chuck Schaff have all been part of the stories I tell here. Without these people I would have had no reservoir of experience on which to draw, nor would I have had as much fun along the way. Members of my laboratory at the University of Chicago—Randall Dahn, Marcus Davis, Adam Franssen, Andrew Gillis, Christian Kammerer, Kalliopi Monoyios, and Becky Shearman—all influenced my thinking and tolerated my time away from the bench as I wrote.

Colleagues who gave their time to provide needed background or comments on the manuscript include Kamla Ahluwalia, Sean Carroll, Michael Coates, Randall Dahn, Marcus Davis, Anna DiRienzo, Andrew Gillis, Lance Grande, Elizabeth Grove, Nicholas Hatsopoulos, Robert Ho, Betty Katsaros, Michael LaBarbera, Chris Lowe, Daniel Margoliash, Kalliopi Monoyios, Jonathan Pritchard, Vicky Prince, Cliff Ragsdale, Nino Ramirez, Callum Ross, Avi Stopper, Cliff Tabin, and John Zeller. Haytham Abu-Zayd helped with many administrative matters. My own teachers of anatomy in the Harvard–MIT Health Sciences and Technology program, Farish A. Jenkins, Jr., and Lee Gehrke, stimulated an interest that has lasted over twenty years.

Key advice at the inception of the project, and inspiration throughout, came from Sean Carroll and Carl Zimmer.

The Wellfleet Public Library (Wellfleet, Massachusetts) provided a comfortable home, and much-needed retreat, where I wrote significant parts of the book. A brief stint at the American Academy in Berlin put me in an environment that proved critical when I was completing the manuscript.

My two bosses, Dr. James Madara, M.D. (CEO, University of

Chicago Medical Center, Vice President for Medical Affairs, Dean and Sara and Harold Thompson Distinguished Service Professor in the Biological Sciences Division and the Pritzker School of Medicine), and John McCarter, Jr. (CEO, The Field Museum), supported this project and the research behind it. It has been a true pleasure to work with such insightful and compassionate leaders.

I have been fortunate to teach at the University of Chicago and to have had the opportunity to interact with the leadership of the Pritzker School of Medicine there. The deans, Holly Humphrey and Halina Bruckner, graciously welcomed a paleontologist to their team. Through interacting with them I came to appreciate the challenges and importance of basic medical education.

It has been a great pleasure to be associated with The Field Museum in Chicago, where I have had the opportunity to work with a unique group of people dedicated to scientific discovery, application, and outreach. These colleagues include Elizabeth Babcock, Joseph Brennan, Sheila Cawley, Jim Croft, Lance Grande, Melissa Hilton, Ed Horner, Debra Moskovits, Laura Sadler, Sean VanDerziel, and Diane White. I am also grateful for the support, guidance, and encouragement I have received from the leaders of the Committee on Science of the Board of Trustees at The Field Museum, James L. Alexander and Adele S. Simmons.

I am indebted to my agent, Katinka Matson, for helping me turn an idea into a proposal and for advice throughout the process. I feel privileged to have worked with Marty Asher, my editor. Like a patient teacher, he gave me a nurturing combination of advice, time, and encouragement to help me find my way. Zachary Wagman contributed to this project in countless ways by being free with his time, keen editorial eye, and good counsel. Dan Frank made insightful suggestions that stimulated me to think about the story in new ways. Jolanta Benal copyedited the text and improved the presentation immeasurably. I am very grateful to Ellen Feldman, Kristen Bearse, and the production team for their hard work under a tight schedule.

My parents, Gloria and Seymour Shubin, always knew that I would write a book, even before I did. Without their faith in me, I doubt that I ever would have put a word on paper.

My wife, Michele Seidl, and our children, Nathaniel and Hannah, have been living with fish—both *Tiktaalik* and this book—for the better part of two years. Michele read and commented on every draft of this text and supported long weekend absences while I wrote. Her patience and love made it all possible.

INDEX

Page numbers in *italics* refer to illustrations.

PENGUIN SCIENCE

THE ORIGIN OF LIFE PAUL DAVIES

Paul Davies presents evidence that life began billions of years ago, arguing that it may well have started on Mars and spread to Earth in rocks blasted off the Red Planet by asteroid impacts. This solution to the riddle of life's origin has sweeping implications for the nature of the universe and our place within it, and opens the way to a radical rethinking of where we came from.

'The best science writer on either side of the Atlantic' *Washington Times*

THE BLIND WATCHMAKER RICHARD DAWKINS

Science is how we know what we are, where we are and why we are. The title of this work refers to the Rev. William Paley's 1802 work, *Natural Theology*, which argued that just as finding a watch would lead you to conclude that a watchmaker must exist, so the complexity of living organisms proves that a Creator exists. Not so, says Dawkins: 'The only watchmaker in nature is the blind forces of physics, deployed in a very special way...it is the blind watchmaker.'

'Mr. Dawkins succeeds admirably in showing how natural selection allows biologists to dispense with such notions as purpose and design, and he does so in a manner readily intelligible to the modern reader' *The New York Times*

UNWEAVING THE RAINBOW RICHARD DAWKINS

Why do poets and artists so often disparage science in their work? Why does so much scientific literature compare poorly with, say, the phone book? Richard Dawkins has taken a wide-ranging view of the subjects of meaning and beauty in this examination of science, mysticism and human nature.

'The product of a beguiling and fascinating mind and one generous enough to attempt to include all willing readers in its brilliantly informed enthusiasm' Melvyn Bragg, *Observer*

IF YOU ENJOYED THIS EXTRACT, WHY NOT TRY THE WHOLE BOOK?

THE STUFF OF THOUGHT:
LANGUAGE AS A WINDOW INTO HUMAN NATURE
STEVEN PINKER

'Moments of genuine revelation and some very good jokes'
Mark Haddon, *Sunday Telegraph*

The Stuff of Thought is an exhilarating work of non-fiction. Surprising, thought-provoking and incredibly enjoyable, there is no other book like it – Steven Pinker will revolutionise the way you think about language. He analyses what words actually mean and how we use them, and he reveals what this can tell us about ourselves. He shows how we use space and motion as metaphors for more abstract ideas, and uncovers the deeper structures of human thought that have been shaped by evolutionary history. He also explores the emotional impact of language, from names to swear words, and shows us the full power that it can have over us. And, with this book, he also shows just how stimulating and entertaining language can be.

'Astonishingly readable' *Daily Telegraph*

'Perceptive, amusing and intelligent' *The Times*

'No one writes about language as clearly as Steven Pinker, and this is his best book yet' *Financial Times*

Get 20% off the retail price of *The Stuff of Thought*.

Go to www.penguin.co.uk/pinker and use the coupon code 'sevenwords' in the shopping basket.

This offer is valid until the 4th of March, 2009.

PENGUIN SCIENCE

THE BLANK SLATE STEVEN PINKER

'The best book on human nature that I or anyone else will ever read. Truly magnificent' Matt Ridley, *Sunday Telegraph*

'A passionate defence of the enduring power of human nature … both life-affirming and deeply satisfying' Tim Lott, *Daily Telegraph*

'Brilliant … enjoyable, informative, clear, humane' *New Scientist*

'If you think the nature/nurture debate has been resolved, you are wrong. It is about to be reignited with a vengeance … this book is required reading' *Literary Review*

'Startling … Pinker makes his main argument persuasively and with great verve … This is a breath of air for a topic that has been politicized for too long' *Economist*

HOW THE MIND WORKS STEVEN PINKER

'Why do memories fade? Why do we lose our tempers? Why do fools fall in love? Pinker's objective in this erudite account is to explore the nature and history of the human mind' *Sunday Times*

'Witty popular science that you enjoy reading for the writing as well as for the science' *The New York Review of Books*

THE LANGUAGE INSTINCT STEVEN PINKER

'A marvellously readable book…illuminates every facet of human language: its biological origin, its uniqueness to humanity, its acquisition by children, its grammatical structure, the production and perception of speech, the pathology of language disorders and its unstoppable evolution' *Nature*

'An extremely valuable book, informative and well written' Noam Chomsky

'Brilliant … Pinker describes every aspect of language, from the resolution of ambiguity to the way speech evolved … he expounds difficult ideas with clarity, wit and polish' Stuart Sutherland, *Observer*

PENGUIN SCIENCE

FREEDOM EVOLVES
DANIEL C. DENNETT

'If anyone can claim to be the Leonardo of the New Renaissance, Dennett can'
Sunday Times

'This is a serious book with a brilliant message' Matt Ridley, *Sunday Telegraph*

'Dennett has produced the most powerful and ingenious attempt at reconciling
Darwinism with the belief in human freedom to date' John Gray, *Independent*

'An outstandingly good book. There is no better philosophical exponent of what
evolutionary biology really means' *The Times*

DARWIN'S DANGEROUS IDEA
DANIEL C. DENNETT

In one of the finest explanations of the nature and implications of Darwinian
evolution ever written, Dennett moves skilfully from a firm foundation in biology
to possible applications for the theory in engineering and cultural evolution,
presenting an unsurpassed analysis of the objections to evolutionary theory along
the way. Extremely lucid, wonderfully written, and scientifically and
philosophically impeccable.

'Dennett's dangerous idea: to use his gift for lucid explanation and his twinkling
wit to cure the strange allergy to Darwin in modern intellectual life. It is essential
– and pleasurable – reading for any thinking person' Steven Pinker

'If you want an exciting, wide-ranging romp through great ideas, read this book.
Daniel Dennett proves fully worthy of Charles Darwin' Jared Diamond

RICHARD P. FEYNMAN

Feynman's clear, uncomplicated thinking and exuberant experiences of life are a joy to read and Penguin is releasing his works in a new series to celebrate the coolest Nobel laureate ever to play the bongos.

'He is everything you want and expect a scientist to be: charming, sceptical, funny, blindingly intelligent ... confirms one's suspicion that Feynman was probably the coolest scientist who ever lived.' *Guardian*

DON'T YOU HAVE TIME TO THINK?

Richard Feynman was no ordinary genius. Brilliant, free-spirited and irreverent, he upset those in authority, gave captivating lectures, wrote equations on napkins in strip joints and touched countless lives everywhere. He also wrote hundreds of witty, eccentric and moving letters to his family, friends, critics, colleagues and devoted fans around the world. These letters are now brought together in this volume for the first time, and will introduce you to a unique person whose wisdom and lust for life inspired all those who came into his orbit.

THE MEANING OF IT ALL

At the peak of his career, maverick genius Richard Feynman gave three public lectures addressing the questions that most inspired and troubled him. Covering everything from the atomic bomb to ethics, the imagination to the meaning of life, they are brought together in this provocative and hugely entertaining volume.

THE PLEASURE OF FINDING THINGS OUT

This collection of the best short works of rule-breaking genius Richard Feynman shows his passion for knowledge and sense of fun at their most infectious. The revealing and inspiring pieces here span a lifetime of enthusiasm for discovering what makes the world tick – including uproarious tales of early student experiments; safecracking and outwitting US censors during the Second World War; his first lecture as a graduate student (to an audience including Albert Einstein); and the memories of the father who delighted in showing him the world and sparked his insatiable curiosity.

PENGUIN PSYCHOLOGY

BLINK
THE POWER OF THINKING WITHOUT THINKING
MALCOLM GLADWELL

'Astonishing … *Blink* really does make you rethink the way you think' *Daily Mail*

'Trust my snap judgement, buy this book: you'll be delighted' *The New York Times*

An art expert sees a ten-million-dollar sculpture and instantly spots it's a fake. A marriage analyst knows within minutes whether a couple will stay together. A fire-fighter suddenly senses he has to get out of a blazing building. A speed dater clicks with the right person …

This book is all about those moments when we 'know' something without knowing why. Here Malcolm Gladwell, one of the world's most original thinkers, explores the phenomenon of 'blink', showing how a snap judgement can be far more effective than a cautious decision. By trusting your instincts, he reveals, you'll never think about thinking in the same way again …

'Compelling, fiendishly clever' *Evening Standard*

'Brilliant … the implications for business, let alone love, are vast' *Observer*

'Superb … this wonderful book should be compulsory reading' *New Statesman*

'*Blink* might just change your life' *Esquire*

'Should you buy this book? You already know the answer to that' *Independent on Sunday*

He just wanted a decent book to read ...

Not too much to ask, is it? It was in 1935 when Allen Lane, Managing Director of Bodley Head Publishers, stood on a platform at Exeter railway station looking for something good to read on his journey back to London. His choice was limited to popular magazines and poor-quality paperbacks – the same choice faced every day by the vast majority of readers, few of whom could afford hardbacks. Lane's disappointment and subsequent anger at the range of books generally available led him to found a company – and change the world.

'We believed in the existence in this country of a vast reading public for intelligent books at a low price, and staked everything on it'
Sir Allen Lane, 1902–1970, founder of Penguin Books

The quality paperback had arrived – and not just in bookshops. Lane was adamant that his Penguins should appear in chain stores and tobacconists, and should cost no more than a packet of cigarettes.

Reading habits (and cigarette prices) have changed since 1935, but Penguin still believes in publishing the best books for everybody to enjoy. We still believe that good design costs no more than bad design, and we still believe that quality books published passionately and responsibly make the world a better place.

So wherever you see the little bird – whether it's on a piece of prize-winning literary fiction or a celebrity autobiography, political tour de force or historical masterpiece, a serial-killer thriller, reference book, world classic or a piece of pure escapism – you can bet that it represents the very best that the genre has to offer.

Whatever you like to read – trust Penguin.